Methods in Molecular Biology

Series Editor
John M. Walker
School of Life and Medical Sciences
University of Hertfordshire
Hatfield, Hertfordshire, UK

For further volumes:
http://www.springer.com/series/7651

For over 35 years, biological scientists have come to rely on the research protocols and methodologies in the critically acclaimed *Methods in Molecular Biology* series. The series was the first to introduce the step-by-step protocols approach that has become the standard in all biomedical protocol publishing. Each protocol is provided in readily-reproducible step-by-step fashion, opening with an introductory overview, a list of the materials and reagents needed to complete the experiment, and followed by a detailed procedure that is supported with a helpful notes section offering tips and tricks of the trade as well as troubleshooting advice. These hallmark features were introduced by series editor Dr. John Walker and constitute the key ingredient in each and every volume of the *Methods in Molecular Biology* series. Tested and trusted, comprehensive and reliable, all protocols from the series are indexed in PubMed.

Nitrogen Metabolism in Plants

Methods and Protocols

Edited by

Kapuganti Jagadis Gupta

National Institute of Plant Genome Research, New Delhi, India

Humana Press

Editor
Kapuganti Jagadis Gupta
National Institute of Plant Genome
Research
New Delhi, India

ISSN 1064-3745 ISSN 1940-6029 (electronic)
Methods in Molecular Biology
ISBN 978-1-4939-9789-3 ISBN 978-1-4939-9790-9 (eBook)
https://doi.org/10.1007/978-1-4939-9790-9

This Humana imprint is published by the registered company Springer Science+Business Media, LLC, part of Springer Nature.
The registered company address is: 233 Spring Street, New York, NY 10013, U.S.A.

Preface

All living organisms require nitrogen. It is an essential element and is placed next to carbon for its requirement and great value throughout the plant life cycle. It is an important and crucial constituent of various amino acids, proteins, nucleotides, chlorophyll, cofactors, enzymes, vitamins, secondary metabolites, and growth hormones. It should be noted, that although nitrogen is present/abundant in the atmosphere, plant can't utilize it directly. Nitrogen has to be converted first to an usable form for its assimilation to occur. In this context, nitrogen cycle plays an important role in interconversion of various nitrogen forms. For instance, lightening, ammonification, nitrification, denitrification, nitric oxide oxidation, nitrate reduction, nitrite reduction, NOx emissions, bacterial nitrogen fixation via symbiosis, and free living organisms play a role in the operation of nitrogen cycle.

Since nitrogen (N) is important for plant growth. N deficiency can cause damage to growth, photosynthetic and reproductive ability, can ultimately affect the yield stability, and can also greatly affect the plant resistance against several biotic and abiotic stresses. The response of plants to utilize the supplied nitrogen should be taken into consideration while determining the amount of fertilizer applied. Several parameters, such as volatilization, denitrification, leaching, and soil erosion, play role in reducing or depleting N levels in soil. In the current situation of N deficiency, the use of synthetic nitrogen fertilizers has dramatically increased to meet N demand for crop productivity. But the increasing N fertilizer application can also increase nitrogen oxide emissions that can primarily cause ozone pollution and contribute towards increase of green house gases. Increasing nitrogen use efficiency (NUE) can greatly improve plant productivity and fight against stress conditions and reduce NOx emissions. Generations of cultivars with high NUE through breeding approaches and genome editing can also greatly aid in sustainable plant productivity. Leguminous plants have evolved mechanisms to associate with nitrogen-fixing bacteria. The fixation of atmospheric nitrogen into ammonia by legumes helps them to obtain sufficient nitrogen for their growth and development. Studying nitrogen metabolism is essential for understanding plant growth, development and stress. In this Methods in Molecular Biology series, several aspects of nitrogen metabolism are covered. This book covers topics on nitrogen metabolism ranging from nitrogen uptake to assimilation.

Plant nitrogen assimilation involves a series of steps. Once nitrate is absorbed by the roots, it is transported to other parts of the plant via well-efficient transport system. Once nitrate is transported inside the cell, nitrate reductase converts nitrate to nitrite in cytosol, and subsequently, the produced nitrite is reduced to ammonia by nitrite reductase present in chloroplasts or plastids. Then, the produced ammonia is assimilated in to various pathways via GS/GOGAT cycle. The important enzymes, such as nitrate reductase, nitrite reductase, glutamine synthase, GOGAT, glutamate dehydrogenase, and alanine amino transferase, are explained in the chapter by Kishorekumar et al. Nitrate reductase is a crucial enzyme. It plays a role in the reduction of nitrate to nitrite. The same enzyme also plays a role in the synthesis of nitric oxide, which is an important signal molecule involved in various plant growth and developmental processes. Wany et al.'s chapter describes the measurement of NR activity, nitrite, and NO. NR becomes active under certain conditions such as low pH, hypoxia, light, etc. The measurement of NR under these conditions is well explained by Bulle et al. NO reacts with protein and causes a modification called *S*-nitrosylation, which takes place when

NO reacts with the sulfhydryl group (–SH) of cysteine residues of proteins or peptides. This reaction leads to the production of *S*-nitrosothiols (SNOs). This NO-mediated PTM is known to play an important role in signaling. The measurement of *S*-nitrosothiols is explained by Mioto et al. *S*-nitrosoglutathione reductase (GSNOR) is a key enzyme regulating intracellular levels of *S*-nitrosoglutathione (GSNO) and protein *S*-nitrosothiols. Janku et al. using spectrophotometric measurements and activity staining in native polyacrylamide gels, describe the measurement of GSNOR in plants. The measurement of gene expression is important to study the regulation of nitrogen metabolism; Bulle et al.'s chapter describes detailed procedures for the measurement of various genes and their modulation under stress conditions. Determination of activities of glutamate dehydrogenase and of the GABA shunt enzymes as the markers under elevated CO_2, temperature increase, flooding, and other stresses. The detailed procedures are explained in the chapter by Eprintsev et al.

Nitrate and ammonium form of nutrition plays a different role in pathogen resistance. Nitrate form of nutrition helps in plant defense via rapid hypersensitive response, whereas under ammonium nutrition, the HR is slow and plants show susceptibility. Studying HR requires the measurement of bacterial growth, ROS, NO production, electrolyte leakage, etc., which are described in Singh et al.'s chapter. Arora et al. describes foldscope methodology to observe superoxide and cell death via trypan blue staining. Using differential nitrate or ammonium feeding, it was shown how the N nutrition affects superoxide production and cell death using foldscope.

The measurement of nitrogen content is crucial for various studies associated in nitrogen metabolism. Singh et al.'s chapter explains detailed methodology to detect the N content in plants and microbes, while Bandyopadhyay and Prasad's chapters explain detailed methods to measure nitrogen use efficiency in foxtail millet (*Setaria italica*) in a pot-based system under different nitrogen nutrition regimes. Studying nitrogen fixation and metabolism requires the isolation and characterization of root nodule bacteria (RNB). These are well explained in the Tak et al.'s chapter. Soil nitrification and denitrification play a role in determining N levels in soil; hence, the measurement of gene expression is very important. Pereira and Filho discuss detailed procedures for gene expression analysis of nitrifying bacteria. Proteins are important parts of nitrogen metabolism. Proteomics play an important role in studying nitrogen metabolism. The extracellular matrix proteins play an essential role in several metabolic processes, including carbon and nitrogen metabolism. The process to perform proteomics of ECM is explained by Elagamey et al.

I am very grateful to the scientists from nine countries who have contributed to the methods described in this book. I extend my gratitude to John Walker for his continuous guidance. I also thank our current Director, Ramesh Sonti, for his continuous support and encouragement, Mallesham and Aprajita for the formatting, Aakanksha for the index, Kishorekumar for the cover page preparation, and Werner Kaiser for introducing me to nitrogen metabolism. Furthermore, I thank my mother and my young son, Rithwik, for their patience during the preparation of this book and my almighty Shirdi Sai Baba for his blessings throughout my journey.

New Delhi, India ***Kapuganti Jagadis Gupta***

Contents

Contributors

REENA ARORA • *National Institute of Plant Genome Research, Aruna Asaf Ali Marg, New Delhi, India*
TIRTHANKAR BANDYOPADHYAY • *National Institute of Plant Genome Research, New Delhi, India*
GARIMA BISSA • *BNF and Microbial Genomics Lab., Department of Botany, Center of Advanced Study, Jai Narain Vyas University, Jodhpur, Rajasthan, India*
MALLESHAM BULLE • *National Institute of Plant Genome Research, Aruna Asaf Ali Marg, New Delhi, India*
NIRANJAN CHAKRABORTY • *National Institute of Plant Genome Research, New Delhi, India*
SUBHRA CHAKRABORTY • *National Institute of Plant Genome Research, New Delhi, India*
FRANCISCO J. CORPAS • *Department of Biochemistry, Cell and Molecular Biology of Plants, Estación Experimental del Zaidín, CSIC, Granada, Spain*
EMAN ELAGAMEY • *National Institute of Plant Genome Research, New Delhi, India; Plant Pathology Research Institute, Agricultural Research Center (ARC), Giza, Egypt*
ALEXANDER T. EPRINTSEV • *Department of Biochemistry and Cell Physiology, Voronezh State University, Voronezh, Russia*
LUCIANO FRESCHI • *Department of Botany, Institute of Biosciences, University of São Paulo, São Paulo, Brazil*
HUKAM S. GEHLOT • *BNF and Microbial Genomics Lab., Department of Botany, Center of Advanced Study, Jai Narain Vyas University, Jodhpur, Rajasthan, India*
KAPUGANTI JAGADIS GUPTA • *National Institute of Plant Genome Research, New Delhi, India*
ABIR U. IGAMBERDIEV • *Department of Biology, Memorial University of Newfoundland, St. John's, NL, Canada*
MARTINA JANKŮ • *Department of Biochemistry, Faculty of Science, Palacký University, Olomouc, Czech Republic*
REDDY KISHOREKUMAR • *National Institute of Plant Genome Research, New Delhi, India*
APRAJITA KUMARI • *National Institute of Plant Genome Research, Aruna Asaf Ali Marg, New Delhi, India*
HAI-BI LI • *Key Laboratory of Sugarcane Biotechnology and Genetic Improvement (Guangxi), Ministry of Agriculture, Sugarcane Research Center, Chinese Academy of Agricultural Sciences, Guangxi key Laboratory of Sugarcane Genetic Improvement, Sugarcane Research Institute, Guangxi Academy of Agricultural Sciences, Nanning, Guangxi, China; College of Agriculture, Guangxi University, Nanning, Guangxi, China*
YANG-RUI LI • *Key Laboratory of Sugarcane Biotechnology and Genetic Improvement (Guangxi), Ministry of Agriculture, Sugarcane Research Center, Chinese Academy of Agricultural Sciences, Guangxi key Laboratory of Sugarcane Genetic Improvement, Sugarcane Research Institute, Guangxi Academy of Agricultural Sciences, Nanning, Guangxi, China; College of Agriculture, Guangxi University, Nanning, Guangxi, China*
LENKA LUHOVÁ • *Department of Biochemistry, Faculty of Science, Palacký University, Olomouc, Czech Republic*
ALEJANDRA MATIZ • *Department of Botany, Institute of Biosciences, University of São Paulo, São Paulo, Brazil*

PAULO T. MIOTO • *Department of Botany, Biological Sciences Center, Universidade Federal de Santa Catarina, Florianópolis, Brazil*
KANIKA NARULA • *National Institute of Plant Genome Research, New Delhi, India*
PRADEEP KUMAR PATHAK • *National Institute of Plant Genome Research, New Delhi, India*
ENGIL I. P. PEREIRA • *School of Earth, Environmental, and Marine Sciences, The University of Texas Rio Grande Valley, Edinburg, TX, USA*
MAREK PETŘIVALSKÝ • *Department of Biochemistry, Faculty of Science, Palacký University, Olomouc, Czech Republic*
MANOJ PRASAD • *National Institute of Plant Genome Research, New Delhi, India*
NATALIA V. SELIVANOVA • *Department of Biochemistry and Cell Physiology, Voronezh State University, Voronezh, Russia*
NAMRATA SINGH • *National Institute of Plant Genome Research, Aruna Asaf Ali Marg, New Delhi, India*
POOJA SINGH • *National Institute of Plant Genome Research, Aruna Asaf Ali Marg, New Delhi, India*
PRATIKSHA SINGH • *Key Laboratory of Sugarcane Biotechnology and Genetic Improvement (Guangxi), Ministry of Agriculture, Sugarcane Research Center, Chinese Academy of Agricultural Sciences, Guangxi key Laboratory of Sugarcane Genetic Improvement, Sugarcane Research Institute, Guangxi Academy of Agricultural Sciences, Nanning, Guangxi, China; College of Agriculture, Guangxi University, Nanning, Guangxi, China; Guangxi key Laboratory of Crop Genetic Improvement and Biotechnology, Nanning, Guangxi, China*
RAJESH KUMAR SINGH • *Key Laboratory of Sugarcane Biotechnology and Genetic Improvement (Guangxi), Ministry of Agriculture, Sugarcane Research Center, Chinese Academy of Agricultural Sciences, Guangxi key Laboratory of Sugarcane Genetic Improvement, Sugarcane Research Institute, Guangxi Academy of Agricultural Sciences, Nanning, Guangxi, China; College of Agriculture, Guangxi University, Nanning, Guangxi, China; Guangxi key Laboratory of Crop Genetic Improvement and Biotechnology, Nanning, Guangxi, China*
QI-QI SONG • *Key Laboratory of Sugarcane Biotechnology and Genetic Improvement (Guangxi), Ministry of Agriculture, Sugarcane Research Center, Chinese Academy of Agricultural Sciences, Guangxi key Laboratory of Sugarcane Genetic Improvement, Sugarcane Research Institute, Guangxi Academy of Agricultural Sciences, Nanning, Guangxi, China; College of Agriculture, Guangxi University, Nanning, Guangxi, China*
NISHA TAK • *BNF and Microbial Genomics Lab., Department of Botany, Center of Advanced Study, Jai Narain Vyas University, Jodhpur, Rajasthan, India*
MARCELO C. M. TEIXEIRA FILHO • *Department of Plant Protection, Rural Engineering and Soils, São Paulo State University (UNESP), Ilha Solteira, SP, Brazil*
TEREZA TICHÁ • *Department of Biochemistry, Faculty of Science, Palacký University, Olomouc, Czech Republic*
AAKANKSHA WANY • *National Institute of Plant Genome Research, Aruna Asaf Ali Marg, New Delhi, India*
LI-TAO YANG • *Key Laboratory of Sugarcane Biotechnology and Genetic Improvement (Guangxi), Ministry of Agriculture, Sugarcane Research Center, Chinese Academy of Agricultural Sciences, Guangxi key Laboratory of Sugarcane Genetic Improvement, Sugarcane Research Institute, Guangxi Academy of Agricultural Sciences, Nanning, Guangxi, China; College of Agriculture, Guangxi University, Nanning, Guangxi, China*

Chapter 1

An Overview of Important Enzymes Involved in Nitrogen Assimilation of Plants

Reddy Kishorekumar, Mallesham Bulle, Aakanksha Wany, and Kapuganti Jagadis Gupta

Abstract

Nitrogen (N) is a macro-nutrient that is essential for growth development and resistance against biotic and abiotic stresses of plants. Nitrogen is a constituent of amino acids, proteins, nucleic acids, chlorophyll, and various primary and secondary metabolites. The atmosphere contains huge amounts of nitrogen but it cannot be taken up directly by plants. Plants can take up nitrogen in the form of nitrate, ammonium, urea, nitrite, or a combination of all these forms. In addition, in various leguminous rhizobia, bacteria can convert atmospheric nitrogen to ammonia and supply it to the plants. The form of nitrogen nutrition is also important in plant growth and resistance against pathogens. Nitrogen content has an important function in crop yield. Nitrogen deficiency can cause reduced root growth, change in root architecture, reduced plant biomass, and reduced photosynthesis. Hence, understanding the function and regulation of N metabolism is important. Several enzymes and intermediates are involved in nitrogen assimilation. Here we provide an overview of the important enzymes such as nitrate reductase, nitrite reductase, glutamine synthase, GOGAT, glutamate dehydrogenase, and alanine aminotransferase that are involved in nitrogen metabolism.

Key words Nitrogen, Nitrate, Ammonium, Glutamate, Glutamine, Alanine

1 Introduction

Plants require nitrogen (N) for their growth, development, and resistance against biotic and abiotic stresses. Nitrogen is important for the biosynthesis of amino acids, proteins, nucleic acids, chlorophyll, various primary and secondary metabolites, and hormones [1]. Most plants assimilate nitrogen in the form of nitrate, which is supplied in the form of inorganic and organic fertilizers [2]. Plants also have the capacity to directly take up supplied ammonia. In addition, leguminous plants are associated with symbiotic bacteria that can fix atmospheric nitrogen to ammonia in specialized structures called nodules. Plant nitrogen assimilation involves various steps. Once nitrate has been absorbed, it is transported to other

Kapuganti Jagadis Gupta (ed.), *Nitrogen Metabolism in Plants: Methods and Protocols*, Methods in Molecular Biology, vol. 2057, https://doi.org/10.1007/978-1-4939-9790-9_1,

parts of the plant via various high- and low-affinity nitrate transporters (HATs, LATs) [3]. When nitrate has been transported to the cell, it is converted to nitrite by nitrate reductase in the cytosol. The produced nitrite is further reduced to ammonia by nitrite reductase in chloroplasts or plastids. Then, the produced ammonia is assimilated to various pathways via the glutamine synthetase/glutamine oxoglutarate aminotransferase (GS/GOGAT) cycle [1, 4]. Ammonium generated via photorespiration and protein degradation pathways is also reassimilated via the GS/GOGAT cycle [1] In this chapter, we provide a detailed overview of nitrate reductase, glutamine synthase, glutamate synthase, glutamate dehydrogenase, and alanine amino transferase.

2 Nitrate Reductase

Nitrate reductase (NR; 1.6.6.1) is an important enzyme involved in N metabolism. This enzyme is located in the cytosol. The major reaction of NR is to convert nitrate to nitrite, which is then further reduced to ammonium in plastids by nitrite reductase (NiR; EC 1.7.7.1) [5]. The NR reaction requires nicotinamide adenine dinucleotide (reduced) (NADH). NR has high affinity toward nitrite. The K_m value for nitrite is 100 μM. In the model plant *Arabidopsis*, NR is encoded by two homologous genes (*NIA1* and *NIA2*) [6]. Recent evidence suggests that NR uses nitrite as a substrate and produces nitric oxide (NO) [7, 8]. Several lines of evidence suggest that NO originated by NR participates in several developmental and physiological functions [7, 9, 10]: antioxidant metabolism [11, 12], stomatal opening [13, 14], floral transition [15], lateral root formation, root development [16, 17], osmotic stress response [18, 19], cold stress [20], and regulation of respiration [12].

NR is a homodimer of 100–115 kDa [21], with a specific flavin adenine dinucleotide (FAD), heme, and the molybdenum co-factor (MoCo) [22–24]. NR exists in two forms: first, NADH-specific NR (EC 1.6.6.1), which is the most common form present in crops, and, in addition to this NADH-specific NR, in maize, barley, rice, and soybean a second form, NAD(P)H-bi-specific NR (1.6.6.2), is present [25].

Several internal (cellular) and external (environmental) cues are involved in the regulation of NR. The steady state of NR protein levels depends on its gene expression and posttranslational modifications. Environmental factors such as soil nitrate and ammonium concentration, light, CO_2, low and high temperature, hypoxia, and salinity also influence NR induction [26–31]. The external supply of nitrate induces NR and nitrate response-related gene expression, and ammonium suppresses its activity and expression [12, 31, 32]. Castaings et al. [33] reported that NR is also regulated by NODULE-INCEPTION-like proteins (NLPs) transcription factors. This regulation helps in modulation of genes involved in nitrate sensing including NR and nitrite and nitrate transporters. Light and dark dynamics is also important in the regulation of NR. For instance, light-induced regulation of NR gene expression in turn mediates diurnal regulation of the phytochromes, which can become crucial to modulate NR during light–dark transition in maize [27]. Transgenic tobacco plants with 35S-*NIA2*-induced *NR* gene expression revealed that NR is regulated by light, thus providing evidence that NR is regulated by translational and post-translational levels [34]. In addition to light, NR activity also depends on CO_2 availability [26, 31]. Ammonium and tungstate also inhibit NR activity [31]. Anoxia under dark conditions promotes the activation of NR and NO production in leaves and roots [31, 35]. Experimental evidence suggests that energy and carbohydrate status influence *NR* gene expression as well as NR activity [36].

Phytohormones such as abscisic acid (ABA) and cytokinin negatively regulate the *NR* transcript levels in barley [37]. Very recently, it has been elucidated in *Arabidopsis*, that cell-cycle gene expression modulates cell-wall integrity in a cytokinin- and nitrate reductase-dependent manner [38]. Salinity differentially regulates NR activity in leaves and roots. Under salt conditions (150 mM), NR activity increased in maize roots, in contrast to leaves, where it was downregulated as much as fourfold [29]. Plant-derived protein hydrolysates (PH) (amino acids and peptides) stimulates N metabolism by regulating key enzymes involved in N assimilation in tomato, including NR [39].

NR is regulated by posttranslational levels as well. A rapid and reversible interconversion of active and inactive states of the protein decides the fate of the enzyme active state (Fig. 1). The posttranslational modifications also modulate NR protein. NR is phosphorylated at the 543 position of spinach [40] in the hinge 1 region, which leads to formation of a binding site for 14-3-3 proteins (R-X-X-pS/pT-X-P).

It was reported that Ca^{2+}-dependent (calmodulin domain) protein kinases (CDPK) activate NR by phosphorylating at ser-534. This position is bound by nitrate reductase inhibitor protein (NIP), that is, 14-3-3, leading to its inactivation [40–42].

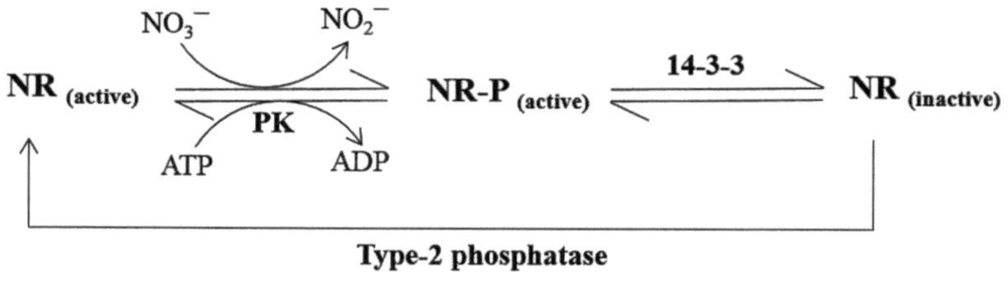

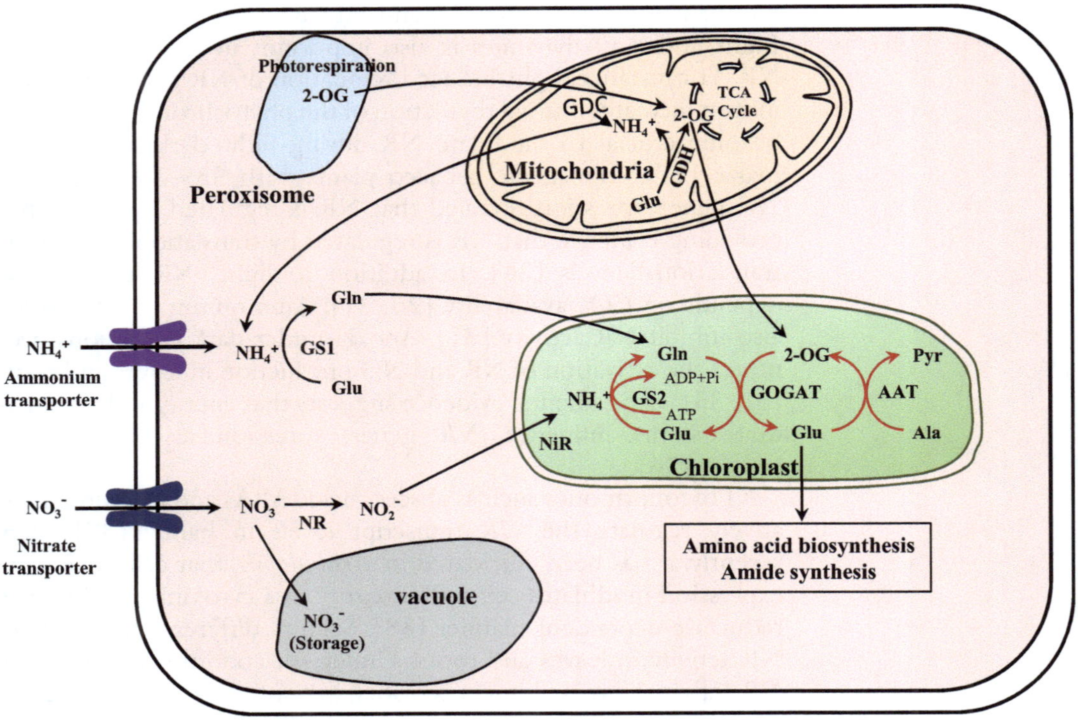

Fig. 1 Sequence of steps involved in plant nitrogen metabolism. Plants take up nitrate (NO_3^-) and ammonium (NH_4^+) ions through nitrate and ammonium transporters. Nitrate is reduced to nitrite (NO_2^-) via nitrate reductase (NR), and the NO_2^- is further reduced to NH_4^+ by nitrite reductase (NiR) in the chloroplasts. Excess NO_3^- is mobilized to vacuoles for storage and osmoregulation. The resulting NH_4^+ is utilized for amino acid biosynthesis and other anabolic reactions via chloroplastic ammonium transporters and alanine transaminase (AAT) reactions. Additionally, the directly transported ammonium from soil via ammonium transporters is incorporated into the glutamine synthetase/glutamine oxoglutarate aminotransferase (GS/GOGAT) cycle in chloroplasts. Additionally NH_4^+ liberated from glutamate dehydrogenase (GDH) and glycine decarboxylase (GDC) is readily utilized by cytosolic GS, followed by the GOGAT cycle

3 Glutamine Synthetase

The reduced ammonium from nitrate reduction in chloroplasts is incorporated into amino acids through the glutamine synthetase (GS) reaction [43]. GS (EC 6.3.1.2) is considered to be one of the key enzymes involved in N metabolism [44]. This enzyme catalyzes synthesis of glutamine from glutamate and ammonia and ATP.

The overall reaction catalyzed by glutamine synthetase is

$$\text{L-glutamate} + \text{ATP} + \text{NH}_3 \rightarrow \text{L-glutamine} + \text{ADP} + \text{Pi} + \text{H}_2\text{O}$$

This enzyme uses ammonia that is derived from various metabolic pathways such as amino acid catabolism, the phenylpropanoid pathway, and photorespiration, and also from ammonia fixed by the rhizobia in nodules [2]. Glutamine is present in very high amounts in xylem and phloem sap and is involved in nitrogen mobilization [45]; it also controls proline in the phloem [46]. Glutamine can be used for biosynthesis of various amino acids, intermediates in the tricarboxylic acid (TCA) cycle, in chlorophylls, and also acts as a nitrogen source for transamination reactions of various metabolic intermediates [47]. This enzyme is involved in nitrogen assimilation via the GS/GOGAT pathway (main ammonium assimilation pathway in plants), which is composed of glutamine synthetase (GS), glutamate synthase (GOGAT), and glutamate dehydrogenase (GDH).

GS is involved in plant growth and grain filling, or yield in cereals [48]. The activity of GS together with total N, chlorophyll, and soluble protein can be considered as indicators for N assimilation and recycling [49]. GS also regulates levels of auxin and root growth [50]. GS exists in two forms, cytosolic (GS1) and plastidial (GS2) [51]. GS1, encoded by three to five nuclear genes [52], is involved in multiple pathways including ammonium transport, seed germination and control of proline production in phloem in addition to primary ammonium assimilation and nitrogen use efficiency [46, 53–55]. In maize, the five cytosolic GS genes that encode GS1 are differentially expressed in the roots, stems, and leaves [56, 57]. GS1 is a heteromeric protein consisting of three subunits of similar MW (38–40 kDa). The multigene nature causes the relative expression and tissue specificity of the *GS1* gene to vary among different species.

The plastidial GS (GS2), encoded by *Gln2*, only expresses photosynthetic tissues [58]. This gene is encoded by a single nuclear gene [59]. In addition to chloroplast localization, the mitochondrial localization of GS2 has been seen in *Arabidopsis* [59, 60]. GS2 assimilates ammonia produced by the photorespiratory pathway and nitrate reduction in the mesophyll cells [61]. The level of *GS2* becomes low during senescence; in such conditions *GS1* becomes the principal enzyme to meet the cellular requirements associated with ammonium assimilation [62]. Recently, the third isoform of GS (GS3) has been identified in wheat leaf [63]. GS3 is homologous to GS2 at 42.1 kDa. The amino acid sequence and subcellular localization are the same as GS2.

4 Glutamate Synthase (GOGAT)

Glutamate synthase (GS) is also called glutamate 2-oxoglutarate aminotransferase (GOGAT). GOGAT is part of the GS/GOGAT cycle, which is involved in the primary route of nitrogen assimilation (Fig. 2). GOGAT transfers the amide group from glutamine to 2-oxoglutarate to produce glutamate.

$$\text{Glutamine} + \text{2-oxoglutarate} + 2\ \text{Fd (red)} \rightarrow 2\ \text{glutamate} + 2\ \text{Fd (oxi)}$$

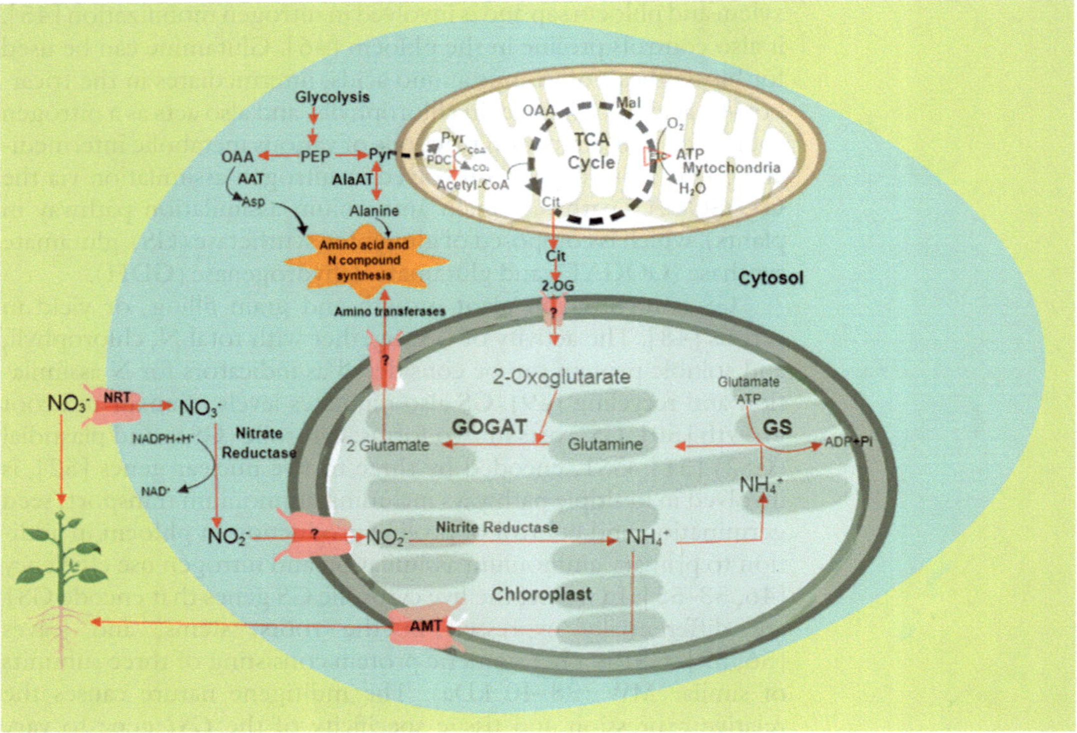

Fig. 2 Integrated view of the GS/GOGAT pathway operation in plants. Ammonium is generated in plastids via NiR or ammonium directly absorbed from soil via NH_4^+, then feeds into the GS/GOGAT cycle of amino acid biosynthesis, where the enzyme glutamine synthetase (GS) converts glutamate and ammonium to glutamine. Glutamine oxoglutarate aminotransferase (GOGAT) then generates glutamate from glutamine produced via the TCA cycle product 2-oxoglutarate (2-OG). Further, glutamate is incorporated into amino acids by aminotransferases. On the other hand, aspartate aminotransferase (AAT) and alanine amino transferase (AlaAT) catalyze the reversible conversion of alanine and α-oxoglutarate to pyruvate and glutamate in cytosol

The synthesized glutamate can be used for subsequent glutamine synthetase catalysis or to form other nitrogen compounds. In higher plants, GOGAT is available in two isoforms, ferredoxin-dependent GOGAT (Fd-GOGAT; EC 1.2.7.1) and NADH-dependent GOGAT (NADH-GOGAT; EC 1.4.1.14): these forms differ in their structure and function [64, 65]. Fd-GOGAT is an iron–sulfur-containing monomeric flavoprotein with a molecular mass of 165 kDa. This enzyme is abundant in plastids and catalyzes the assimilation of ammonium (NH_4^+) generated during photorespiration and the light-dependent reduction of nitrate (NO_3^-) [65]. This enzyme, which accounts for more than 96% of total glutamate synthase activity, is located in the chloroplasts [66].

$$\text{Glutamine} + \text{2-oxoglutarate} + 2\,\text{Fd (red)} \rightarrow 2\,\text{glutamate} + 2\,\text{Fd (oxi)}$$
$$\text{Glutamine} + \text{2-oxoglutarate} + \text{NADH} \rightarrow 2\,\text{glutamate} + \text{NAD}$$

In *Arabidopsis*, Fd-GOGAT is encoded by two different functional genes, *GLU1* and *GLU2*. The *GLU1* gene is expressed in leaves, whereas expression of *GLU2* is more predominant in roots and etiolated tissues [67, 68]. Light positively regulates *GLU1* and *GLU2* gene induction; however, *GLU2* gene induction is regulated by sucrose in the dark [66, 68].

NADH-GOGAT is also an iron–sulfur flavoprotein with a subunit molecular mass of 225–230 kDa. It is found in nonphotosynthetic tissues and predominantly in root nodules of legumes such as alfalfa [64]. It catalyzes the assimilation of NH_4^+ imported from the nitrogen-fixing bacteroides and then incorporated into amino acids. Increased NADH-GOGAT gene expression is proportional to nitrogen fixation and NH_4^+ content.

5 Glutamate Dehydrogenase

NAD-dependent glutamate dehydrogenase (NAD-GDH; EC 1.4.1.2) catalyzes the deamination of glutamate to produce 2-oxoglutarate and generates ammonia (NH_3) and NADH in a reversible reaction:

$$\text{L-glutamate} + \text{NAD(P)} + H_2O \rightarrow \text{2-oxyglutarate} + NH_3 + \text{NAD(P)H} + \text{Pi} + H_2O$$

This reaction takes place in the mitochondria and is associated with the tricarboxylic acid (TCA) cycle. This reaction occurs under low energy conditions (C limited) and amide synthesis [69–71]. This reaction also occurs in conditions such as salt stress [72] or hypoxia [73].

In *Arabidopsis*, GDH is encoded by two nuclear genes, *GDH1* and *GDH2*, composed of α- and β-subunits [74]. These subunits associate differently and form seven active isozymes. All the isoforms may have similar metabolic functions, as they showed similar activity ratios and kinetic properties in grapevine leaf crude extracts [75]. *GDH* isoforms function in both vegetative and reproductive stages during plant development [76]. The third *GDH* gene, *GDH3*, which codes for the γ-subunit, was identified in *Arabidopsis* a decade later, and it is involved in the formation of heteromeric complexes along with the α- and β-subunits [77, 78]. This enzyme is localized in root companion cells [78]. *GDH* is prominently localized in the mitochondria [79]. In conditions of high ammonium accumulation, this enzyme also occurs in the cytosol [80]. Mitochondrial expression of *GDH* is associated with low N availability, whereas its cytosolic counterpart is correlated with ammonium beyond threshold levels [80].

Tobacco transgenic plants overexpressing *GDHA* and *GDHB* genes from *Nicotiana plumbaginifolia* exhibited additional GDH activity, increased leafy starch, enhanced metabolites of carbon and nitrogen metabolism, and accumulation of digalactosylglycerol, erythronate, and porphyrin, consequently contributing better yield and stress tolerance [81]. Overexpression of *Escherichia coli* NADPH-*GDH* in tobacco plants showed higher biomass as well as exhibiting increased herbicidal resistance [82, 83]. *E. coli* NADPH-*GDH* overexpression in maize caused increased grain biomass under drought conditions [84]. Metabolic profiling of GDH overexpressers showed increased accumulation of di-galactosylglycerol (DGG), erythronate, porphyrin, and lower aspartate content. These metabolites are known to function under salt stress [85, 86] and temperature stress [87].

Dark and sucrose-starved conditions induce the expression of *GDH* to increase energy efficiency [74]. *GDH3* is regulated by N starvation and cytokinin treatment in *Arabidopsis* [76]. Anoxia and reoxygenation conditions increase *GDH1* and *GDH2* activity in an ethylene-dependent manner [88]. Using the *GDH* triple mutant (*gdh1-2-3*) lacking GDH activity, it was demonstrated that GHD is involved in 2-oxoglutarate (2-OG) synthesis [89].

6 Alanine Aminotransferase

Alanine aminotransferase (*AlaAT*; E.C.2.6.1.2) is an important enzyme in nitrogen metabolism. This pyridoxal phosphate-dependent enzyme, distributed in various plant tissues and organs [90], catalyzes the reversible transfer of an amino group from glutamate to pyruvate to form 2-oxoglutarate and alanine [91]:

$$\text{Pyruvate} + \text{glutamate} \rightleftharpoons \text{Alanine Trasaminase} + \text{2-oxyglutarate}$$

AlaAT activity is found in leaves, roots, flowers, and endosperm [92, 93] and has been characterized in different plant species such as soybean (*Glycine max*), legumes (*Medicago truncatula*), canola (*Brassica napus*), *Arabidopsis* (*Arabidopsis thaliana*), rice (*Oryza sativa*), and *Lotus japonicas* [90, 91, 94–97]. Studies on the *AlaAT* gene family have mainly focused on *Arabidopsis*. In *A. thaliana*, four *AlaAT* genes have been identified [98]. Previously, it was shown that hypoxia treatment leads to increased expression of *AlaAT1* and *AlaAT2* in *Arabidopsis*, and treatment has also increased AlaT activity and increased alanine without any harmful consequences [91]. The primary role of AlaT is to remove excess alanine. Moreover, in a T-DNA knockout mutant of *Arabidopsis* plants, alaat1-1 produced alanine despite reduced AlaAT activity [91]. Later, it was found that *AlaAT* expression is regulated by hypoxia, light, and differential N levels [99, 100]. AlaAT is critical in rapid conversion of alanine to pyruvate during recovery from hypoxic stress [91]. Plants overexpressing *AlaAT* have increased biomass and yield and consequently improved N use efficiency (NUE) at low N supply (0.5–4 mM) [96, 101]. Similarly, transgenic sugarcane plants overexpressing barley *AlaAT* showed improved NUE compared with untransformed plants under low N conditions [102]. In *Lotus japonicas,* AlaAT is linked with glycolysis and the TCA cycle during hypoxia induced by waterlogging, thus contributing to hypoxic stress tolerance [103]. It was demonstrated that the high levels of *AlaAT* during reoxygenation following the hypoxia phase drive rapid conversion of accumulated alanine back into glutamate, suggesting that *AlaAT* also has a function during the recovery phase [103].

References

1. Krapp A (2005) Plant nitrogen assimilation and its regulation: a complex puzzle with missing pieces. Curr Opin Plant Biol 25:115–122
2. Temple SJ, Vance CP, Gantt JS (1998) Glutamate synthase and nitrogen assimilation. Trends Plant Sci 3:51–56
3. Crawford NM, Glass AD (1998) Molecular and physiological aspects of nitrate uptake in plants. Trends Plant Sci 3:389–395
4. O'Brien JA, Vega A, Bouguyon E, Krouk G, Gojon A, Coruzzi G, Gutiérrez RA (2016) Nitrate transport, sensing, and responses in plants. Mol Plant 9:837–856
5. Oaks A (1994) Primary nitrogen assimilation in higher plants and its regulation. Can J Bot 72:739–750
6. Wilkinson JQ, Crawford NM (1993) Identification and characterization of a chlorate-resistant mutant of *Arabidopsis thaliana* with mutations in both nitrate reductase structural genes NIA1 and NIA2. Mol Gen Genet 239:289–297
7. Gupta KJ, Fernie AR, Kaiser WM, van Dongen JT (2011) On the origins of nitric oxide. Trends Plant Sci 16:160–168
8. Rockel P, Strube F, Rockel A, Wildt J, Kaiser WM (2002) Regulation of nitric oxide (NO) production by plant nitrate reductase in vivo and in vitro. J Exp Bot 53:103–110
9. Astier J, Gross I, Durner J (2018) Nitric oxide production in plants: an update. J Exp Bot 69:3401–3411
10. Mur LA, Mandon J, Persijn S, Cristescu SM, Moshkov IE, Novikova GV, Hall MA, Harren FJ, Hebelstrup KH, Gupta KJ (2013) Nitric oxide in plants: an assessment of the current state of knowledge. AoB Plants 5
11. Sang J, Jiang M, Lin F, Xu S, Zhang A, Tan M (2008) Nitric oxide reduces hydrogen peroxide accumulation involved in water stress-induced subcellular anti-oxidant defense in maize plants. J Integr Plant Biol 50:231–243
12. Wany A, Gupta AK, Kumari A, Mishra S, Singh N, Pandey S, Vanvari R, Igamberdiev AU, Fernie AR, Gupta KJ (2018) Nitrate nutrition influences multiple factors in order to increase energy efficiency under hypoxia in *Arabidopsis*. Ann Bot 123(4):691–705. https://doi.org/10.1093/aob/mcy202
13. Desikan R, Griffiths R, Hancock J, Neill S (2002) A new role for an old enzyme: nitrate reductase-mediated nitric oxide generation is required for abscisic acid-induced stomatal closure in *Arabidopsis thaliana*. Proc Natl Acad Sci USA 99:16314–16318
14. Hao F, Zhao S, Dong H, Zhang H, Sun L, Miao C (2010) Nia1 and Nia2 are involved in exogenous salicylic acid-induced nitric oxide generation and stomatal closure in *Arabidopsis*. J Integr Plant Biol 52:298–307
15. Seligman K, Saviani EE, Oliveira HC, Pinto-Maglio CAF, Salgado I (2008) Floral transition and nitric oxide emission during flower development in *Arabidopsis thaliana* is affected in nitrate reductase-deficient plants. Plant Cell Physiol 49:1112–1121
16. Kolbert Z, Bartha B, Erdei L (2008) Exogenous auxin-induced NO synthesis is nitrate reductase-associated in *Arabidopsis thaliana* root primordia. J Plant Physiol 165:967–975
17. Lombardo MC, Lamattina L (2012) Nitric oxide is essential for vesicle formation and trafficking in *Arabidopsis* root hair growth. J Exp Bot 63:4875–4885
18. Kolbert Z, Ortega L, Erdei L (2010) Involvement of nitrate reductase (NR) in osmotic stress-induced NO generation of *Arabidopsis thaliana* L. roots. J Plant Physiol 167:77–80

19. Royo B, Moran JF, Ratcliffe RG, Gupta KJ (2015) Nitric oxide induces the alternative oxidase pathway in *Arabidopsis* seedlings deprived of inorganic phosphate. J Exp Bot 66:6273–6280
20. Zhao MG, Chen L, Zhang LL, Zhang WH (2009) Nitric reductase-dependent nitric oxide production is involved in cold acclimation and freezing tolerance in *Arabidopsis*. Plant Physiol 151:755–767
21. Solomonson LP, Barber MJ (1990) Assimilatory nitrate reductase: functional properties and regulation. Annu Rev Plant Physiol Plant Mol Biol 41:225–253
22. Crawford NM, Smith M, Bellissimo D, Davis RW (1988) Sequence and nitrate regulation of the *Arabidopsis thaliana* mRNA encoding nitrate reductase, a metalloflavoprotein with three functional domains. Proc Natl Acad Sci USA 85:5006–5010
23. Neame PJ, Barber MJ (1989) Conserved domains in molybdenum hydroxylases. The amino acid sequence of chicken hepatic sulfite oxidase. J Biol Chem 264:20894–20901
24. Campbell WH (2001) Structure and function of eukaryotic NAD(P)H:nitrate reductase. Cell Mol Life Sci 58:194–204
25. Kleinhofs A, Warner RL (1990) Advances in nitrate assimilation. In: Miflin BJ, Lea PJ (eds) The biochemistry of plants, Intermediary nitrogen metabolism, vol 16. Academic Press, San Diego, pp 89–120
26. Kaiser WM, Brendle-Behnisch E (1991) Rapid modulation of spinach leaf nitrate reductase activity by photosynthesis: I. Modulation in vivo by CO_2 availability. Plant Physiol 96:363–367
27. Bowsher CG, Long DM, Oaks A, Rothstein SJ (1991) Effect of light/dark cycles on expression of nitrate assimilatory genes in maize shoots and roots. Plant Physiol 95:281–285
28. Crawford NM (1995) Nitrate: nutrient and signal for plant growth. Plant Cell 7:859–868
29. Baki GAE, Siefritz F, Man HM, Weiner H, Kaldenhoff R, Kaiser WM (2000) Nitrate reductase in *Zea mays* L. under salinity. Plant Cell Environ 23:515–521
30. Campbell WH (1996) Nitrate reductase biochemistry comes of age. Plant Physiol 111:355–361
31. Planchet E, Jagadis Gupta K, Sonoda M, Kaiser WM (2005) Nitric oxide emission from tobacco leaves and cell suspensions: rate limiting factors and evidence for the involvement of mitochondrial electron transport. Plant J 41:732–743
32. Neyra CA, Hageman RH (1975) Nitrate uptake and induction of nitrate reductase in excised corn roots. Plant Physiol 56:692–695
33. Castaings L, Camargo A, Pocholle D, Gaudon V, Texier Y, Boutet-Mercey S, Meyer C (2009) The nodule inception-like protein 7 modulates nitrate sensing and metabolism in *Arabidopsis*. Plant J 57:426–435
34. Vincentz M, Caboche M (1991) Constitutive expression of nitrate reductase allows normal growth and development of *Nicotiana plumbaginifolia* plants. EMBO J 10:1027–1035
35. Gupta KJ, Stoimenova M, Kaiser WM (2005) In higher plants, only root mitochondria, but not leaf mitochondria, reduce nitrite to NO, in vitro and in situ. J Exp Bot 56:2601–2609
36. Botrel A, Kaiser WM (1997) Nitrate reductase activation state in barley roots in relation to the energy and carbohydrate status. Planta (Berl) 201:496–501
37. Hoff T, Truong HN, Caboche M (1994) The use of mutants and transgenic plants to study nitrate assimilation. Plant Cell Environ 17:489–506
38. Gigli-Bisceglia N, Engelsdorf T, Strnad M, Vaahtera L, Khan GA, Jamoune A, Hamann T (2018) Cell wall integrity modulates *Arabidopsis thaliana* cell cycle gene expression in a cytokinin- and nitrate reductase-dependent manner. Development (Camb) 145:145(19). dev166678. https://doi.org/10.1242/dev.166678
39. Sestili F, Rouphael Y, Cardarelli M, Pucci A, Bonini P, Canaguier R, Colla G (2018) Protein hydrolysate stimulates growth in tomato coupled with N-dependent gene expression involved in N assimilation. Front Plant Sci 9:1233
40. Bachmann M, Shiraishi N, Campbell WH, Yoo BC, Harmon AC, Huber SC (1996) Identification of Ser-543 as the major regulatory phosphorylation site in spinach leaf nitrate reductase. Plant Cell 8:505–517
41. Kaiser WM, Spill D (1991) Rapid modulation of spinach leaf nitrate reductase by photosynthesis: II. In vitro modulation by ATP and AMP. Plant Physiol 96:368–375
42. Douglas P, Moorhead G, Hong Y, Morrice N, MacKintosh C (1998) Purification of a nitrate reductase kinase from *Spinacea oleracea* leaves, and its identification as a calmodulin-domain protein kinase. Planta (Berl) 206:435–442
43. Elliott WH (1953) Isolation of glutamine synthetase and glutamotransferase from green peas. J Biol Chem 201:661–672

44. Hirel B, Andrieu B, Valadier MH, Renard S, Quillere I, Chelle M, Pommel B, Fournier C, Drouet JL (2005) Physiology of maize II: identification of physiological markers representative of the nitrogen status of maize (*Zea mays*) leaves during grain filling. Physiol Plant 124:178–188
45. Masclaux-Daubresse C, Daniel-Vedele F, Dechorgnat J, Chardon F, Gaufichon L, Suzuki A (2010) Nitrogen uptake, assimilation and remobilization in plants: challenges for sustainable and productive agriculture. Ann Bot 105:1141–1157
46. Brugière N, Dubois F, Limami AM, Lelandais M, Roux Y, Sangwan RS, Hirel B (1999) Glutamine synthetase in the phloem plays a major role in controlling proline production. Plant Cell 11:1995–2011
47. Lea PJ, Miflin BJ (2003) Glutamate synthase and the synthesis of glutamate in plants. Plant Physiol Biochem 41:555–564
48. Moison M, Marmagne A, Dinant S, Soulay F, Azzopardi M, Lothier J, Avice JC (2018) Three cytosolic glutamine synthetase isoforms located in different order veins work together for N remobilization and seed filling in *Arabidopsis*. J Exp Bot 69:4379–4393
49. Martin A, Lee J, Kichey T, Gerentes D, Zivy M, Tatout C, Tercé-Laforgue T (2006) Two cytosolic glutamine synthetase isoforms of maize are specifically involved in the control of grain production. Plant Cell 18:3252–3274
50. Lebedev VG, Korobova AV, Shendel GV, Kudoyarova GR, Shestibratov KA (2018) Effect of glutamine synthetase gene overexpression in birch (*Betula pubescens*) plants on auxin content and rooting in vitro. Dokl Biochem Biophys 48:143–145
51. Hirel B, Lea PJ (2001) Ammonia assimilation. In: Lea PJ, Morot-Gaudry JF (eds) Plant nitrogen. Springer, Berlin, Heidelberg
52. Bernard SM, Møller AL, Dionisio G, Kichey T, Jahn TP, Dubois F, Baudo M, Lopes MS, Tercé-Laforgue T, Foyer CH, Parry MA (2008) Gene expression, cellular localization and function of glutamine synthetase isozymes in wheat (*Triticum aestivum* L.). Plant Mol Biol 67:89–105
53. Ishiyama K, Inoue E, Tabuchi M, Yamaya T, Takahashi H (2004) Biochemical background and compartmentalized functions of cytosolic glutamine synthetase for active ammonium assimilation in rice roots. Plant Cell Physiol 45:1640–1647
54. Guan M, Møller IS, Schjørring JK (2014) Two cytosolic glutamine synthetase isoforms play specific roles for seed germination and seed yield structure in *Arabidopsis*. J Exp Bot 66:203–212
55. Zhang Z, Xiong S, Wei Y, Meng X, Wang X, Ma X (2017) The role of glutamine synthetase isozymes in enhancing nitrogen use efficiency of N-efficient winter wheat. Sci Rep 7:1000
56. Sakakibara H, Shimizu H, Hase T, Yamazaki Y, Takao T, Shimonishi Y, Sugiyama T (1996) Molecular identification and characterization of cytosolic isoforms of glutamine synthetase in maize roots. J Biol Chem 271:29561–29568
57. Li MG, Villemur R, Hussey PJ, Silflow CD, Gantt JS, Snustad DP (1993) Differential expression of six glutamine synthetase genes in *Zea mays*. Plant Mol Biol 23:401–440
58. Guiz C, Hirel B, Shedlofsky G, Gadal P (1979) Occurrence and influence of light on the relative proportions of two glutamine sythetases in rice leaves. Plant Sci Lett 15:271–277
59. Taira M, Valtersson U, Burkhardt B, Ludwig RA (2004) *Arabidopsis thaliana* GLN_2-encoded glutamine synthetase is dual targeted to leaf mitochondria and chloroplasts. Plant Cell 16:2048–2058
60. Mann AF, Fentem PA, Stewart GR (1979) Identification of two forms of glutamine synthetase in barley (*Hordeum vulgare*). Biochem Biophys Res Commun 88:515–521
61. Kumagai E, Araki T, Hamaoka N, Ueno O (2011) Ammonia emission from rice leaves in relation to photorespiration and genotypic differences in glutamine synthetase activity. Ann Bot 108:1381–1386
62. Guo Y, Cai Z, Gan S (2004) Transcriptome of *Arabidopsis* leaf senescence. Plant Cell Environ 27:521–549
63. Wang X, Wei Y, Shi L, Ma X, Theg SM (2015) New isoforms and assembly of glutamine synthetase in the leaf of wheat (*Triticum aestivum* L.). J Exp Bot 66:6827–6834
64. Gregerson RG, Miller SS, Twary SN, Gantt JS, Vance CP (1993) Molecular characterization of NADH-dependent glutamate synthase from alfalfa nodules. Plant Cell 5:215–226
65. Sakakibara H, Watanabe M, Hase T, Sugiyama T (1991) Molecular cloning and characterization of complementary DNA encoding for ferredoxin-dependent glutamate synthase in maize leaf. J Biol Chem 266:2028–2035
66. Suzuki A, Rothstein S (1997) Structure and regulation of ferredoxin-dependent glutamase synthase from *Arabidopsis thaliana*: cloning of cDNA, expression in different

tissues of wild-type and gltS mutant strains, and light induction. Eur J Biochem 243:708–718

67. Lam HM, Coschigano KT, Oliveira IC, Melo-Oliveira R, Coruzzi GM (1996) The molecular-genetics of nitrogen assimilation into amino acids in higher plants. Annu Rev Plant Biol 47:569–593
68. Oliveira IC, Lam HM, Coschigano K, Melo-Oliveira R, Coruzzi G (1997) Molecular-genetic dissection of ammonium assimilation in *Arabidopsis thaliana*. Plant Physiol Biochem 35:185–198
69. Fox GG, Ratcliffe RG, Robinson SA, Stewart GR (1995) Evidence for deamination by glutamate dehydrogenase in higher plants: commentary. Can J Bot 73:1112–1115
70. Purnell MP, Botella JR (2007) Tobacco isoenzyme 1 of NAD (H)-dependent glutamate dehydrogenase catabolizes glutamate in vivo. Plant Physiol 143:530–539
71. Miyashita Y, Good AG (2008) NAD(H)-dependent glutamate dehydrogenase is essential for the survival of *Arabidopsis thaliana* during dark-induced carbon starvation. J Exp Bot 59:667–680
72. Skopelitis DS, Paranychianakis NV, Paschalidis KA, Pliakonis ED, Delis ID, Yakoumakis DI, Kouvarakis A, Papadakis AK, Stephanou EG, Roubelakis-Angelakis KA (2006) Abiotic stress generates ROS that signal expression of anionic glutamate dehydrogenases to form glutamate for proline synthesis in tobacco and grapevine. Plant Cell 18:2767–2781
73. Limami AM, Glévarec G, Ricoult C, Cliquet JB, Planchet E (2008) Concerted modulation of alanine and glutamate metabolism in young *Medicago truncatula* seedlings under hypoxic stress. J Exp Bot 59:2325–2335
74. Melo-Oliveira R, Oliveira IC, Coruzzi GM (1996) *Arabidopsis* mutant analysis and gene regulation define a nonredundant role for glutamate dehydrogenase in nitrogen assimilation. Proc Natl Acad Sci USA 93:4718–4723
75. Loulakakis KA, Roubelakis-Angelakis KA (1996) The seven NAD (H)-glutamate dehydrogenase isoenzymes exhibit similar anabolic and catabolic activities. Physiol Planta 96:29–35
76. Marchi L, Degola F, Polverini E, Tercé-Laforgue T, Dubois F, Hirel B, Restivo FM (2013) Glutamate dehydrogenase isoenzyme 3 (GDH3) of *Arabidopsis thaliana* is regulated by a combined effect of nitrogen and cytokinin. Plant Physiol Biochem 73:368–374
77. Yamada K, Lim J, Dale JM, Chen H, Shinn P, Palm CJ, Southwick AM, Wu HC, Kim C, Nguyen M, Pham P (2003) Empirical analysis of transcriptional activity in the *Arabidopsis* genome. Science 302:842–846
78. Fontaine JX, Tercé-Laforgue T, Bouton S, Pageau K, Lea PJ, Dubois F, Hirel B (2013) Further insights into the isoenzyme composition and activity of glutamate dehydrogenase in *Arabidopsis thaliana*. Plant Signal Behav 8: e23329
79. Dubois F, Tercé-Laforgue T, Gonzalez-Moro MB, Estavillo JM, Sangwan R, Gallais A, Hirel B (2003) Glutamate dehydrogenase in plants: is there a new story for an old enzyme? Plant Physiol Biochem 41:565–576
80. Tercé-Laforgue T, Dubois F, Ferrario-Méry S, de Crecenzo MAP, Sangwan R, Hirel B (2004) Glutamate dehydrogenase of tobacco is mainly induced in the cytosol of phloem companion cells when ammonia is provided either externally or released during photorespiration. Plant Physiol 136:4308–4317
81. Tercé-Laforgue T, Clément G, Marchi L, Restivo FM, Lea PJ, Hirel B (2015) Resolving the role of plant NAD-glutamate dehydrogenase: III. Overexpressing individually or simultaneously the two enzyme subunits under salt stress induces changes in the leaf metabolic profile and increases plant biomass production. Plant Cell Physiol 56:1918–1929
82. Ameziane R, Bernhard K, Lightfoot D (2000) Expression of the bacterial gdhA gene encoding a NADPH glutamate dehydrogenase in tobacco affects plant growth and development. Plant Soil 221:47–57
83. Nolte SA, Young BG, Mungur R, Lightfoot DA (2004) The glutamate dehydrogenase gene gdhA increased the resistance of tobacco to glufosinate. Weed Res 44:335–339
84. Lightfoot DA, Mungur R, Ameziane R, Nolte S, Long L, Bernhard K, Young B (2007) Improved drought tolerance of transgenic *Zea mays* plants that express the glutamate dehydrogenase gene (gdhA) of *E. coli*. Euphytica 156:103–116
85. Kazachkova Y, Batushansky A, Cisneros A, Tel-Zur N, Fait A, Barak S (2013) Growth platform-dependent and independent phenotypic and metabolic responses of *Arabidopsis thaliana* and its halophytic relative *Eutrema salsugineum*, to salt stress. Plant Physiol 162:1583–1598
86. Gechev TS, Hille J, Woerdenbag HJ, Benina M, Mehterov N, Toneva V, Mueller-Roeber B (2014) Natural products from

resurrection plants: potential for medical applications. Biotechnol Adv 32:1091–1101

87. Kaplan F, Kopka J, Haskell DW, Zhao W, Schiller KC, Gatzke N, Guy CL (2004) Exploring the temperature-stress metabolome of *Arabidopsis*. Plant Physiol 136:4159–4168
88. Tsai KJ, Lin CY, Ting CY, Shih MC (2016) Ethylene-regulated glutamate dehydrogenase fine-tunes metabolism during anoxia-reoxygenation. Plant Physiol 172:1548–1562
89. Fontaine JX, Tercé-Laforgue T, Armengaud P, Clément G, Renou JP, Pelletier S, Hirel B (2012) Characterization of a NADH-dependent glutamate dehydrogenase mutant of *Arabidopsis* demonstrates the key role of this enzyme in root carbon and nitrogen metabolism. Plant Cell 24:4044–4065
90. Rocha M, Sodek L, Licausi F, Hameed MW, Dornelas MC, Van Dongen JT (2010) Analysis of alanine aminotransferase in various organs of soybean (*Glycine max*) and in dependence of different nitrogen fertilisers during hypoxic stress. Amino Acids 39:1043–1053
91. Miyashita Y, Dolferus R, Ismond KP, Good AG (2007) Alanine aminotransferase catalyses the breakdown of alanine after hypoxia in *Arabidopsis thaliana*. Plant J 49:1108–1121
92. Kikuchi H, Hirose S, Toki S, Akama K, Takaiwa F (1999) Molecular characterization of a gene for alanine aminotransferase from rice (*Oryza sativa*). Plant Mol Biol 39:149–159
93. Igarashi D, Miwa T, Seki M, Kobayashi M, Kato T, Tabata S, Shinozaki K, Ohsumi C (2003) Identification of photorespiratory glutamate: glyoxylate aminotransferase (GGAT) gene in *Arabidopsis*. Plant J 33:975–987
94. De Sousa CA, Sodek L (2003) Alanine metabolism and alanine aminotransferase activity in soybean (*Glycine max*) during hypoxia of the root system and subsequent return to normoxia. Environ Exp Bot 50:1–8
95. Ricoult C, Echeverria LO, Cliquet JB, Limami AM (2006) Characterization of alanine aminotransferase (AlaAT) multi gene family and hypoxic response in young seedlings of the model legume *Medicago truncatula*. J Exp Bot 57:3079–3089
96. Good AG, Johnson SJ, De Pauw M, Carroll RT, Savidov N, Vidmar J, Lu Z, Taylor G, Stroeher V (2007) Engineering nitrogen use efficiency with alanine amino transferase. Botany 85:252–262
97. Beatty PH, Shrawat AK, Carroll RT, Zhu T, Good AG (2009) Transcriptome analysis of nitrogen-efficient rice over-expressing alanine aminotransferase. Plant Biotechnol J 7:562–576
98. Liepman AH, Olsen LJ (2003) Alanine aminotransferase homologs catalyze the glutamate: glyoxylate aminotransferase reaction in peroxisomes of *Arabidopsis*. Plant Physiol 131:215–227
99. Son D, Kobe A, Sugiyama T (1992) Nitrogen-dependent regulation of the gene for alanine aminotransferase which is involved in the C4 pathway of *Panicum miliaceum*. Plant Cell Physiol 33:507–509
100. Muench DG, Christopher ME, Good AG (1998) Cloning and expression of a hypoxic and nitrogen inducible maize alanine aminotransferase gene. Physiol Plant 103:503–512
101. Shrawat AK, Carroll RT, DePauw M, Taylor GJ, Good AG (2008) Genetic engineering of improved nitrogen use efficiency in rice by the tissue-specific expression of alanine aminotransferase. Plant Biotechnol J 6:722–732
102. Snyman SJ, Hajari E, Watt MP, Lu Y, Kridl JC (2015) Improved nitrogen use efficiency in transgenic sugarcane: phenotypic assessment in a pot trial under low nitrogen conditions. Plant Cell Rep 34:667–669
103. Rocha M, Licausi F, Araujo WL, Nunes-Nesi A, Sodek L, Fernie AR, van Dongen JT (2010) Glycolysis and the tricarboxylic acid cycle are linked by alanine aminotransferase during hypoxia induced by waterlogging of *Lotus japonicus*. Plant Physiol 152:1501–1513

Chapter 2

Methods for Measuring Nitrate Reductase, Nitrite Levels, and Nitric Oxide from Plant Tissues

Aakanksha Wany, Pradeep Kumar Pathak, and Kapuganti Jagadis Gupta

Abstract

Nitrogen (N) is one of the most important nutrients which exist in both inorganic and organic forms. Plants assimilate inorganic form of N [nitrate (NO_3^-), nitrite (NO_2^-) or ammonium (NH_4^+)] and incorporate into amino acids. The metabolism of N involves a series of events such as sensing, uptake, and assimilation. The initial stage is sensing, triggered by nitrate or ammonium signals initiating signal transduction processes in N metabolism. The assimilation pathway initiates with NO_3^-/NH_4^+ transport to roots via specific high and low affinity (HATs and LATs) nitrate transporters or directly via ammonium transporters (AMTs). In cytosol the NO_3^- is reduced to NO_2^- by cytosolic nitrate reductase (NR) and the produced NO_2^- is further reduced to NH_4^+ by nitrite reductase (NiR) in plastids. NR has capability to reduce NO_2^- to nitric oxide (NO) under specific conditions such as hypoxia, low pH, and pathogen infection. The produced NO acts as a signal for wide range of processes such as plant growth development and stress. Here, we provide methods to measure NR activity, NO_2^- levels, and NO production in plant tissues.

Key words DAF-FM, Nitrate, Nitrate reductase, Nitrite, Nitric oxide

1 Introduction

Nitrogen (N) is the most abundant among other gases present in the earth atmosphere which shares almost 78% of total gases. This concentration is very high in comparison to other gases in atmosphere such as oxygen (21%), argon (0.9%), carbon dioxide (0.03%). Despite of this abundance (78%), plants don't possess mechanism to take atmospheric N directly. Plants have the capability to take up N in various forms such as nitrate (NO_3^-), ammonium (NH_4^+) or depending on various factors they also take combination of both NO_3^- and NH_4^+. In most of the agricultural soils and landscapes, NO_3^- is the predominant utilizable form of N for plants than ammonium [1]. Plant growth and developmental processes are impacted by N and it has been a crucial component in contributing towards higher crop productivity and yield stability [2]. N plays a role not only in promoting optimal growth but also

Kapuganti Jagadis Gupta (ed.), *Nitrogen Metabolism in Plants: Methods and Protocols*, Methods in Molecular Biology, vol. 2057, https://doi.org/10.1007/978-1-4939-9790-9_2,

plays a major role in alleviating abiotic and biotic stresses [3]. Thus, it plays a role in sustainable plant productivity under stress [4]. N is also a constituent of many plant secondary metabolites, thus playing an important role in providing tolerance to biotic stresses [5]. In soil, the composition of N vary depending on physiochemical properties of soil and presence of nitrifying, denitrifying microbial activities and other parameters such as soil moisture, oxygen and temperature [6]. Plant metabolism is highly flexible. Plants have capabilities to adapt their metabolism to fluctuating N. These adaptive mechanisms are sensitive and selective enough to facilitate N uptake and assimilation and promote growth under N-limiting conditions, thus playing an important role in achieving nitrogen use efficiency (NUE). Nitrate not only acts as a nutrient, but also an important regulatory signal [7].

The available soil NO_3^- is taken up by several sets of nitrate transporters. The operation of these nitrate transporters (NRTs) is based on their affinity to the available nitrate present in the soil, and these are classified as high affinity (HATs, i.e., *NRT2.1, NRT2.2, NRT2.4*) and low affinity transporters (LATs; *CLCa*, *NRT1.1*, *NPF1.2*) [8, 9]. Along with nitrate transporters, nitrite transporters also play a role in N metabolism [10].

Once nitrate is taken up, the cytosolic NR (NR; 1.6.6.1) catalyzes the reduction of NO_3^- to NO_2^- [Reaction (1)] [11]. This reduction reaction yields NO_2^- and also requires NAD(P)H as an electron donor (Fig. 1b).

$$NO_3^- + NAD(P)H + H^+ + 2e^- \rightarrow NO_2^- + NAD(P)^+ + H_2O \quad (1)$$

The above reaction is considered as a rate limiting step in the N metabolism pathway [12]. The produced NO_2^- is then reduced to NH_4^+ in plastids by plastidial nitrite reductase [Reaction (2)] (NiR; EC 1.7.7.1).

$$NO_2^- + 6_{Fdred} + 8H^+ + 6e^- \rightarrow NH_4^+ + 6_{Fdox} + 2H_2O \quad (2)$$

The activity of NR is directly dependent on various factors such as NO_3^- concentration (K_m ~100 μM) and light conditions [12, 13]. Moreover, several lines of evidence suggest that NR activity is highly induced under hypoxia [12, 14, 15]. Under conditions such as hypoxia/anoxia conditions the cytosolic pH decreases that leads to inhibition of NO_2^- reduction to NH_4^+ (Fig. 1b). These events lead to accumulation of an intermediate nitrite which is one of the limiting factors for the production of NO by NR and subsequent nitrite-NO pathway mediated by mitochondrial electron transport chain [16]. For instance, the NO_2^- levels increases manyfold within several hours of hypoxia treatment in wheat and spinach [11, 12, 17]. Interestingly nitrite acts as an

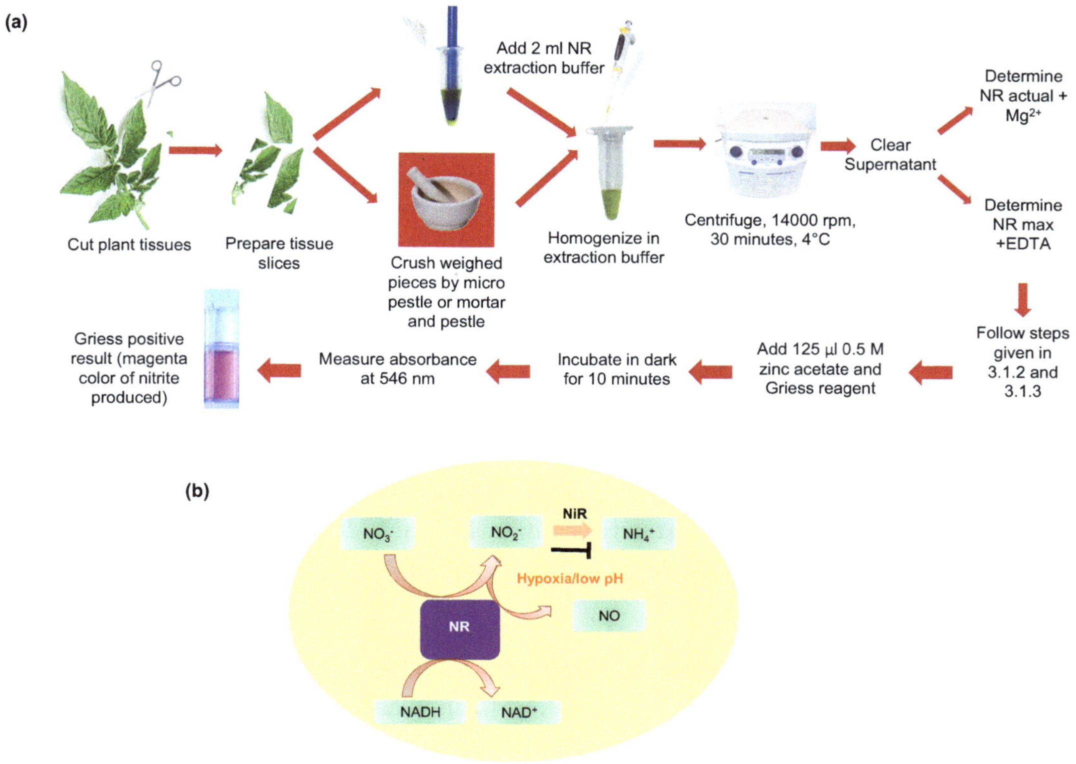

Fig. 1 Nitrate reductase activity measurement (**a**) Steps involved in the NR assay (**b**) Schematic overview of steps involved in nitrate reduction. Cytosolic nitrate reductase (NR) reduces nitrate to nitrite and further reduces to ammonium by utilizing NAD(P)H as an electron donor. Under conditions such as hypoxia or low pH, NO_2^- reduction to NH_4^+ is inhibited, leading to the accumulation of nitrite and subsequent reduction of NO_2^- to NO

inducer of nitrite reductase (NiR) in many plants such as beans [18], Lemna [19], and radish [20]. The nitrite formed can be recycled back to nitrate in plant tissues via phytoglobin nitric oxide cycle [21–25]. Nitrite is also transported to mitochondria where it is reduced to NO by cytochrome c oxidase (Complex IV) and alternative oxidase (AOX) utilizing NAD(P)H and this reaction can produce very low amount of ATP during anoxic conditions for hypoxic survival [25].

The NO generated via NR and mitochondria has been shown to play key roles in plants. For instance, NR derived NO plays role in development, metabolism, seed germination, oxygen homeostasis, root growth, reduction of ROS regulation of respiration, stomatal closure, and adaptive responses to abiotic stresses [26]. In vivo NO measurements can be performed by using fluorescent probes such as DAF-FM DA (4-amino-5-methylamino-2′,7-′-difluorofluorescein diacetate) [14, 15], DAF-2DA (4,5-diaminofluorescein diacetate) and DAR-4 M (diaminorhodamine-4 MAM;

3′,6′-bis(dimethylamino)-9-[2-acetomethoxycarbonyl-3-amino-4-(*N*-methylamino)phenyl]xanthylium iodide [27].

NO-deficient mutants such as *nia1*, *nia2*, or *nia1,2* are often used to elucidate the role of NO. Another alternative approach to diminish NR activity is to use tungstate. This approach has been widely practiced to provide evidence on the role of NR-dependent NO in many biological processes [28]. One should note that tungstate also inhibits various other enzymes in addition to NR. But it is widely used to control NO production in various studies [29, 30]. In order to understand N metabolism, it is essential to measure NR, NO_2^-, and for elucidating cell signaling function NO measurement is needed.

2 Materials

2.1 Nitrate Reductase Activity

2.1.1 Reagents

1. NR extraction buffer: 100 mM HEPES-KOH pH 7.6, 3.5 mM mercaptoethanol, 10 μM FAD, 15 mM $MgCl_2$, 0.5% PVP, 0.5% BSA, and 0.3% of Triton X 100.
2. NR assay buffer I: 100 mM HEPES-KOH pH 7.6, 10 μM FAD, 15 mM $MgCl_2$, 5 mM KNO_3, and 0.2 mM NADH.
3. Stop solution: 0.5 M Zinc acetate.
4. NR assay buffer II: 100 mM HEPES-KOH pH 7.6, 10 μM FAD, and 15 mM EDTA.
5. NED Solution: 0.1% *N*-1-napthylethylenediamine dihydrochloride in water.
6. Sulfanilamide Solution: 1% sulfanilamide in 5% phosphoric acid.

2.1.2 Equipment

1. Centrifuge.
2. Spectrophotometer/Colorimeter.
3. −20 °C freezer.

2.2 Nitrite Measurement

1. NED solution: 0.1% *N*-1-napthylethylenediamine dihydrochloride in water.
2. Sulfanilamide solution: 1% sulfanilamide in 5% phosphoric acid.
3. Nitrite standard: 0.1 M potassium nitrite in water.
4. Extraction buffer: 100 mM HEPES-KOH pH 7.6, 3.5 mM mercaptoethanol, 10 μM FAD, 15 mM $MgCl_2$, 0.5% PVP, 0.5% BSA, and 0.3% of Triton X 100.
5. Activity buffer: 100 mM HEPES-KOH pH 7.6, 10 μM FAD, 15 mM EDTA, 5 mM KNO_3, and 0.2 mM NADH.
6. 0.1 M KNO_2 (potassium nitrite).
7. 20 mM EDTA.

2.3 Nitric Oxide (NO) Measurement Using DAF-FM Diacetate Fluorescent Dyes

2.3.1 Chemicals Required

1. 100 mM HEPES buffer (pH 7.2 to 7.4).
2. 10% Glycerol.
3. DAF-FM dye (10 μM- working concentration).
4. Carboxy- PTIO potassium salt (200 μM- working concentration).
5. Anhydrous dimethyl sulfoxide (DMSO).

2.3.2 Equipment

Fluorescence microscope (Nikon eclipse-*80i*, Japan).

2.4 NO Measurement Using Gas Phase Griess Reagent Assay

2.4.1 Chemicals Required

1. NED solution: 0.2% *N*-1-napthylethylenediamine dihydrochloride in water.
2. Sulfanilamide solution: 2% sulfanilamide in 5% phosphoric acid.

2.4.2 Other Items

Small air pumps.

A surgical syringe with needle.

Petri plates.

Microcentrifuge tubes (0.5 ml).

Nitrogen and oxygen gas cylinders.

3 Methods

3.1 Nitrate Reductase Activity

NR activity can also be performed from tungstate (200–500 μM) grown plants to check the active involvement of NR (Fig. 1a).

3.1.1 Preparation of Extract

1. Excise the plant tissues with scissors and immediately shock freeze grind in liquid nitrogen into a fine powder (1 gFW).
2. Homogenize the fine tissue powder in 2 ml of NR extraction buffer using a micro pestle (Subheading 2.1.1).
3. Centrifuge the homogenate at 10,976 × *g* (maximum speed) for 30 min at 4 °C until it yields a clear supernatant. Repeat this step if needed.
4. Remove the clear supernatant carefully and the aliquots of this supernatant can be directly used for the measurement of NR activity, or store this supernatant at −20 °C until further use.
5. Measure the protein concentration present in the supernatant using Bradford's method [31].

3.1.2 Determination of NR Actual (+ Mg^{2+})

1. Add 100–200 μl of the extracted supernatant into the 800–900 μl of NR assay buffer I (Subheading 2.1.1). Incubate it for 5 min (24 °C) after incubation stop the reaction by adding 100 to 125 μl of zinc acetate (0.5 M).

3.1.3 Determination of Maximal NR (+ EDTA)

1. Transfer 100–200 μl of the supernatant into another tube, and add 20 mM EDTA and incubate at 24 °C for 10 min.
2. After incubation add the NR assay buffer II in 1 ml reaction mixture.
3. Wait for few minutes and then add 5 mM KNO_3 and 0.2 mM NADH (this step will start the reaction again). Incubate this reaction mixture for a period of 10 min.
4. Stop the reaction by adding 100–125 μl zinc acetate (0.5 M) and Griess reagent (0.5 ml sulfanilamide and 0.5 ml NED solution) and incubate it for 10 min in dark at room temperature.
5. Centrifuge at maximum speed 10,976 × *g* for 2 min.
6. Measure the developed color (magenta) using spectrophotometer or colorimeter set at 546 nm.
7. Calibrate the unknown values of nitrite with the nitrite standard reference curve (*see* **Notes 1–9**) to know the concentration of the NR extract.
8. NR activity should be expressed in μmol/mg protein/h.

3.2 Nitrite Measurement

3.2.1 Preparation of Standard Curve

1. Prepare the dilution series of potassium nitrite for the preparing nitrite standard curve (*see* **Notes 3–9**).

3.2.2 Griess Reaction for Nitrite Measurement

1. Place the sulfanilamide solution and NED solution from refrigerator to room temperature in order to equilibrate at room temperature (15–30 min).
2. Add 50 μl of supernatant (extract) to tubes in several replicates.
3. Dispense 50 μl of the sulfanilamide solution to all experimental samples and to tubes/wells containing the dilution series for the nitrite standard reference curve.
4. Incubate 5–10 min at room temperature in dark.
5. Dispense 50 μl of the NED Solution to all tubes/wells.
6. Incubate at room temperature for 5–10 min, in dark. A purple/magenta color will begin to emerge immediately (Fig. 2a).
7. Add 100 to 125 μl of zinc acetate (0.5 M) to each tube/well to stop the reaction.

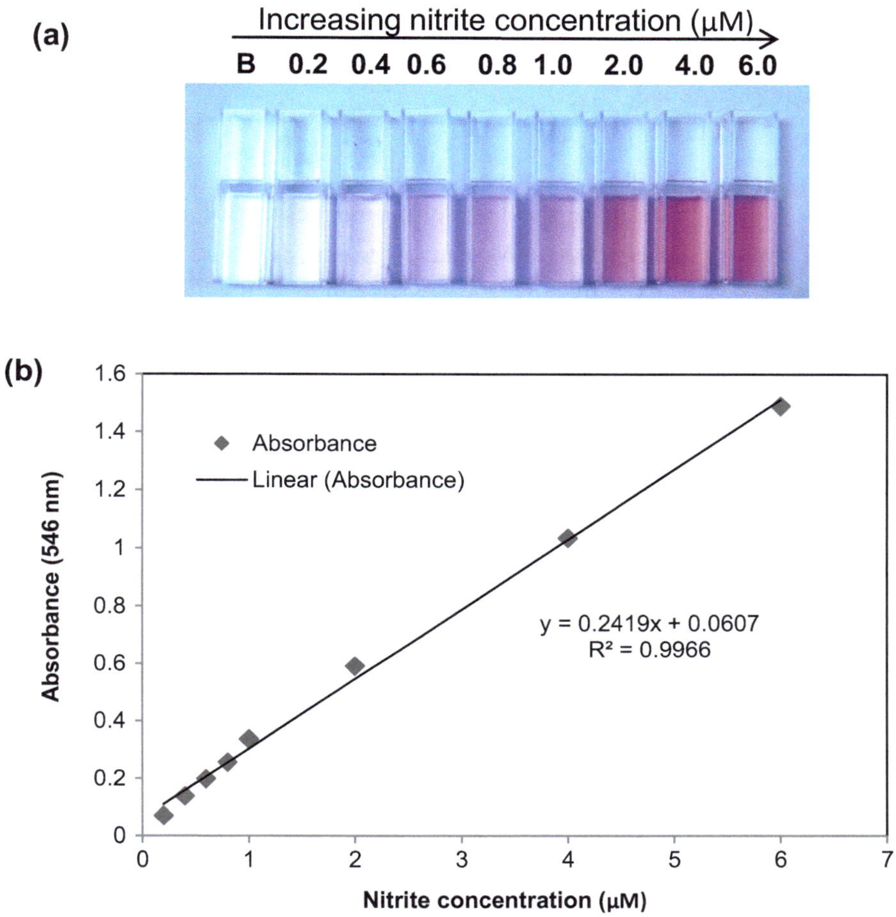

Fig. 2 Standard nitrite curve from 0–6 μM nitrite mixed 1:1 with Griess reagent (NED Solution and Sulfanilamide Solution) (**a**) The magenta color develops when the Griess reagent is added to the nitrite solution. (**b**) A linear standard curve is shown with values ranging from 0–6 μM nitrite

8. Measure absorbance within 30 min in a plate reader or spectrophotometer with a filter set at 546 nm (or a range between 520 nm and 550 nm).

3.2.3 Determination of Nitrite Concentrations in the Extracts

1. To prepare a nitrite standard reference curve, plot the average absorbance value of each concentration of the nitrite standard in *Y* axis with corresponding nitrite concentration as in *X* axis (Fig. 2b).
2. Determine average absorbance value of each experimental sample.
3. Determine the exact concentration by comparing to the nitrite standard reference curve.

3.3 NO Estimation

3.3.1 In Vivo NO Estimation from Roots

1. Carefully excise root segments (2 cm long) from plant seedlings using sharp scalpel and fine forceps. Depending on tissue and requirement of localization cross sections also can be prepared by (10 mm) by free-hand sectioning using a sharp razor blade [14, 15].
2. Place the freshly cut root segments in small petri dishes containing nutrient solution or respective growth media or buffer.
3. For DAF-FM incubation transfer these fine root segments to a 1.5 ml microcentrifuge tube (amber) that contains 100 mM HEPES buffer (pH 7.2) and 10 μM DAF-FM dye. Incubate these tubes for 10 min in dark *see* **Notes 10–15**.
4. After 10 min of incubation wash the root samples three to four times using 100 mM HEPES buffer and transfer the roots onto the slide and mount with a drop of 10% glycerol and cover with cover glass.
5. For cPTIO controls, take another 1.5 ml microcentrifuge tube, add 10 μM DAF-FM and 200 μM of NO scavenger cPTIO to the samples. Incubate, wash and transfer to the slides as same way. cPTIO control act as an internal control to find the observed fluorescence is due to NO or not.
6. Observe the samples in a fluorescence microscope (Nikon, *80i*) with 495 nm excitation and 515 nm emission wavelengths.
7. Quantify the DAF-FM fluorescence using Image J software.

3.3.2 In Vivo NO Estimation from Leaves

1. Select the plants and infiltrate the abaxial side of the leaf (attached to the plant) with a low concentration of DAF-FM DA solution (~approx. 2 μM DAF-FM DA). Incubate the entire plant in dark for 20–30 min. This allows DAF-FM to react with basal NO present in the leaf.
2. After incubation, cut that infiltrated leaf with sharp scalpel and peel off the abaxial side of epidermis very carefully with a sharp sterile forceps having fine tips.
3. Cut the DAF-FM infiltrated area of leaf into (left after peeling epidermis) into small pieces and again incubate these slices 10 μM DAF-DM DA dye for 10 min in dark. This will allow the DAF dye to react with any residual NO produced (whether it is control or treated leaf as per your experiment) *see* **Notes 10–15**.
4. After incubation, transfer the leaf segments in to a small petri dish and wash three to four times using 100 mM HEPES buffer and transfer onto the slide and mount with a drop of 10% glycerol with a cover glass.
5. For cPTIO control follow procedure of Subheading 3.3.1, **step 5**.

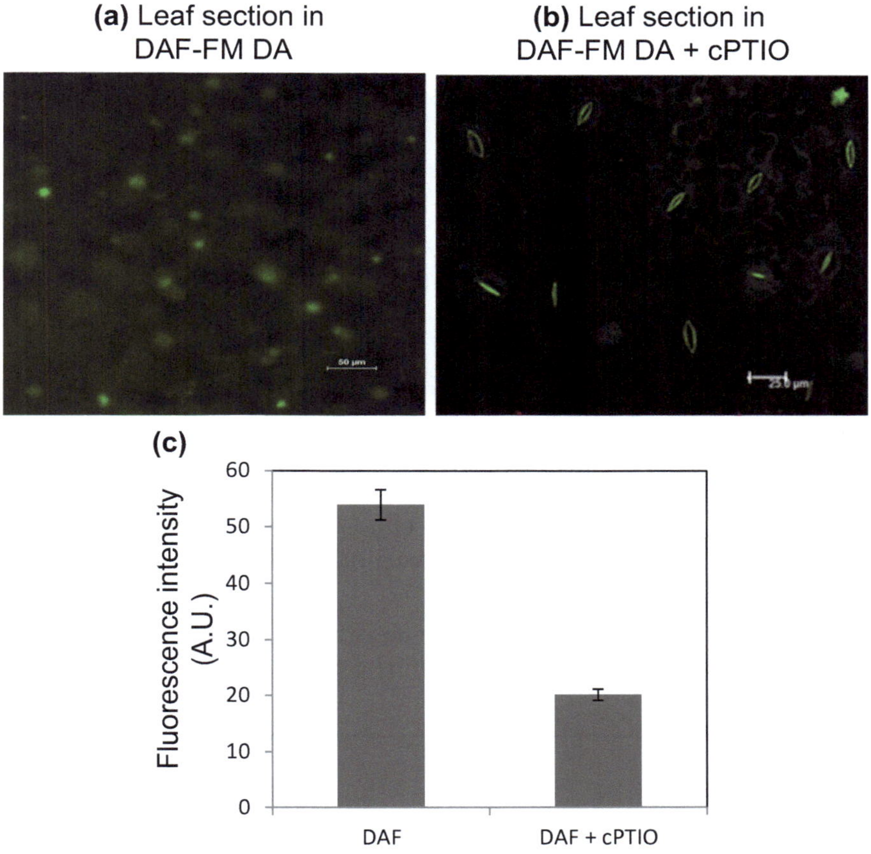

Fig. 3 (**a**) DAF-FM fluorescence in abaxial leaf (peeled epidermis) pathogen infiltrated leaves in tomato (20×, Scale bar-50 μm) with DAF-FM DA, (**b**) Another epidermal section in DAF-FM DA + cPTIO scale bar 25 μm, (**c**) DAF fluorescence intensity in leaves calculated using Image J

6. Observe the samples in a fluorescence microscope (Nikon, 80i) with 495 nm excitation and 515 nm emission wavelengths (Fig. 3a and b).
7. Quantify DAF fluorescence using Image J (Fig. 3c) software.

3.3.3 Gas Phase NO Estimation

1. Create an inlet and outlet valve in customized flasks (Fig. 4). Alternatively prepare 50 ml falcon tubes with inlet and outlet.
2. Put intact plant samples/roots/tissue of interest in the conical flask.
3. Connect the inlet to air pump to flush gas and the outlet to another conical flask containing the Griess reagent (0.2% NED solution and 2% sulfanilamide in 5% phosphoric acid).
4. Depending on measurement requirement flush air or hypoxic gas into the inlet of first conical flask for 1 h or more.

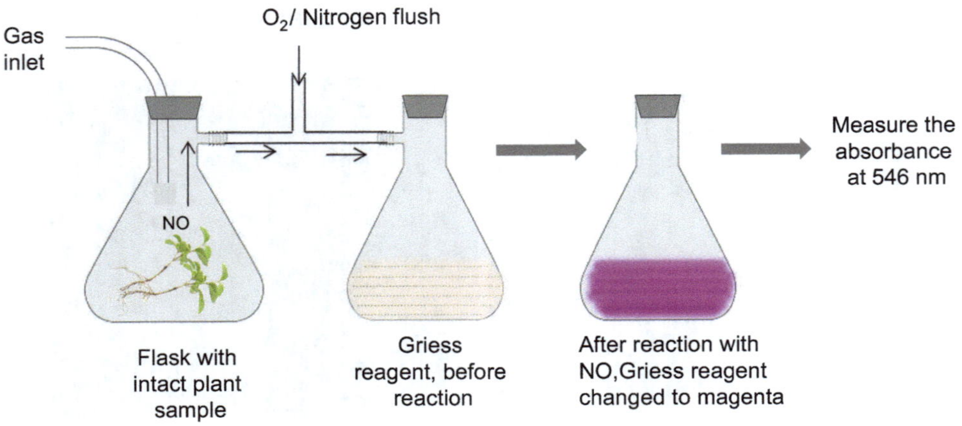

Fig. 4 Gas phase Griess reagent assay to measure NO

5. The color of the Griess reagent changes to purple/magenta due to the NO reaction with Griess reagent, in presence of oxygen.
6. Measure the color absorbance at 540 nm and perform calculations similarly to nitrite estimation.

4 Notes

1. Depending on available equipment NR activity can also be measured in 48 or 96-microwell plates. Individual cuvettes depending upon the number of samples.
2. Cantharidine (a PP2A inhibitor) can be added to prevent dephosphorylation of NR. In the present protocol it is not used, but this step is optional.
3. Nitrite standard reference curve must be prepared freshly for each assay for accurate quantification of NO_2^- levels in experimental samples.
4. Always prepare reference curve(s) of nitrite in the same buffer used for experimental samples for NR activity.
5. Always store Griess reagent components at 4 °C, protected from light.
6. Keep the above solutions back to 4 °C immediately after use.
7. Store Griess reagent components separately; the shelf life is decreased substantially when the reagents are stored as mixed solution.
8. It is important to note that sulfanilamide and NED compete for nitrite in the Griess reaction; the method achieves great sensitivity when the two components are added sequentially.

9. Add the sulfanilamide solution to the sample first, incubate for 5–10 min, then add the NED Solution. This gives better results.
10. DAF-FM diacetate should be stored at −20 °C, and it should be protected from light [32].
11. Prepare DAF-FM stock in high-quality anhydrous DMSO as per the pack size.
12. For long term storage of DAF-FM stocks, distribute the solutions into small aliquots in order to minimize freeze thawing. These solutions can be stable for at least 6 months if stored properly.
13. Working solutions (10 μM) should be prepared immediately before use. The DMSO stock solutions can be easily diluted into aqueous buffers.
14. Before starting the experiment, allow the solutions to equilibrate at room temperature prior to opening.
15. It is recommended that the diluted reagent be not stored for later use.

References

1. Crawford NM, Forde BG (2002) Molecular and developmental biology of inorganic nitrogen nutrition. Arabidopsis Book 1:e0011
2. Landrein B, Formosa-Jordan P, Malivert A, Schuster C, Melnyk CW, Yang W, Turnbull C, Meyerowitz EM, Locke JCW, Jönsson H (2018) Nitrate modulates stem cell dynamics in Arabidopsis shoot meristems through cytokinins. Proc Natl Acad Sci U S A 115:1382–1387
3. Gupta KJ, Brotman Y, Segu S, Zeier T, Zeier J, Persijn ST, Cristescu SM, Harren FJM, Bauwe H, Fernie AR (2013) The form of nitrogen nutrition affects resistance against *Pseudomonas syringae* pv. *phaseolicola* in tobacco. J Exp Bot 64:553–568
4. Wang WX, Vinocur B, Altman A (2003) Plant responses to drought, salinity and extreme temperatures: towards genetic engineering for stress tolerance. Planta 218:1–14
5. Wink M (1999) Functions of plant secondary metabolites and their exploitation in biotechnology. In: Annual plant reviews, vol 3. CRC Press, Boca Raton, FL
6. Schimel JP, Bennett J (2004) Nitrogen mineralization: challenges of a changing paradigm. Ecology 85:591–602
7. Crawford NM (1995) Nitrate: nutrient and signal for plant growth. Plant Cell 7:859–868
8. Krapp A, David LC, Chardin C, Girin T, Marmagne A, Leprince AS, Chaillou S, Ferrario-Méry S, Meyer C, Daniel-Vedele F (2014) Nitrate transport and signalling in arabidopsis. J Exp Bot 65:789–798
9. Tischner T (2000) Nitrate uptake review in higher and lower plants. Plant Cell Environ 23:1005–1024
10. Pao SS, Paulsen IT, Saier MH Jr (1998) Major facilitator superfamily. Microbiol Mol Biol Rev 62:1–34
11. Oaks A (1994) Primary nitrogen assimilation in higher plants and its regulation. Can J Bot 72:739–750
12. Planchet E, Gupta KJ, Sonoda M, Kaiser WM (2005) Nitric oxide emission from tobacco leaves and cell suspensions: rate limiting factors and evidence for the involvement of mitochondrial electron transport. Plant J 41:732–743
13. Hageman RH, Flesher D (1960) Nitrate reductase activity in corn seedlings as affected by light and nitrate content of nutrient media. Plant Physiol 35:700–708
14. Wany A, Kumari A, Gupta KJ (2017) Nitric oxide is essential for the development of aerenchyma in wheat roots under hypoxic stress. Plant Cell Environ 40:3002–3017
15. Wany A, Gupta KJ (2018) Reactive oxygen species, nitric oxide production and antioxidant gene expression during development of

aerenchyma formation in wheat. Plant Signal Behav 13:e1428515

16. Gupta KJ, Igamberdiev AU (2015) Reactive oxygen and nitrogen species signaling and communication in plants. In: Gupta KJ, Igamberdiev AU (eds) Springer, vol 23. ISSN 1867-9048
17. Rockel P, Strube F, Rockel A, Wildt J, Kaiser WM (2002) Regulation of nitric oxide (NO) production by plant nitrate reductase in vivo and in vitro. J Exp Bot 53:103–110
18. Sluiters-ScholtenCMTh (1973) Effect of chloramphenicol and cycloheximide on the induction of nitrate reductase and nitrite reductase in bean leaves. Planta 113:229–240
19. Stewart GR (1972) The regulation of nitrite reductase level in *Lemna minor* L. J Exp Bot 23:171–183
20. Ingle J, Joy KW, Hageman RH (1966) The regulation of activity of the enzymes involved in the assimilation of nitrate by higher plants. Biochem J 100:577–588
21. Aslam M, Rosichan JL, Huffaker RC (1987) Comparative induction of nitrate reductase by nitrate and nitrite in barley leaves. Plant Physiol 83:579–584
22. Funkhouser EA, Garay AS (1981) Appearance of nitrate in soybean seedlings and Chlorella caused by nitrogen starvation. Plant Cell Physiol 22:1279–1286
23. Kaplan D, Roth-Bejerano N, Lips H (1974) Nitrate reductase as a product-inducible enzyme. Eur J Biochem 49:393–398
24. Lips SH, Kaplan D, Roth-Bejerano N (1973) Studies on the induction of nitrate reductase by nitrite in bean-seed cotyledons. Eur J Biochem 37:589–592
25. Stoimenova M, Igamberdiev AU, Gupta KJ, Hill RD (2007) Nitrite-driven anaerobic ATP synthesis in barley and rice root mitochondria. Planta 226:465–474
26. Gupta KJ, Fernie AR, Kaiser WM, van Dongen JT (2011) On the origins of nitric oxide. Trends Plant Sci 16:160–168
27. Planchet E, Kaiser WM (2006) Nitric oxide (NO) detection by DAF fluorescence and chemiluminescence: a comparison using abiotic and biotic NO sources. J Exp Bot 57:3043–3055
28. Xiong J, Fu G, Yang Y, Zhu C, Tao L (2011) Tungstate: is it really a specific nitrate reductase inhibitor in plant nitric oxide research? J Exp Bot 63:33–41
29. Reda M, Golicka A, Kabała K, Janicka M (2018) Involvement of NR and PM-NR in NO biosynthesis in cucumber plants subjected to salt stress. Plant Sci 267:55–64
30. Liang Y, Zheng P, Li S, Li K, Xu H (2018) Nitrate reductase-dependent NO production is involved in H_2S-induced nitrate stress tolerance in tomato via activation of antioxidant enzymes. Sci Hortic 229:207–214
31. Bradford MM (1976) A rapid and sensitive method for the quantitation of microgram quantities of protein utilizing the principle of protein dye binding. Anal Biochem 72:248–254
32. Wany A, Gupta KJ (2016) Localization of nitric oxide in wheat roots by DAF fluorescence. Methods Mol Biol 1424:39–47

Chapter 3

Measurement of Nitrate Reductase Activity in Tomato (*Solanum lycopersicum L.*) Leaves Under Different Conditions

Mallesham Bulle, Reddy Kishorekumar, Pradeep K. Pathak, Aakanksha Wany, and Kapuganti Jagadis Gupta

Abstract

Nitrogen is one of the crucial macronutrients essential for plant growth, development, and survival under stress conditions. Depending on cellular requirement, plants can absorb nitrogen mainly in multiple forms such as nitrate (NO_3^-) or ammonium (NH_4^+) or combination of both via efficient and highly regulated transport systems in roots. In addition, nitrogen-fixing symbiotic bacteria can fix atmospheric nitrogen in to NH_4^+ via highly regulated complex enzyme system and supply to the roots in nodules of several species of leguminous plants. If NO_3^- is a primary source, it is transported from roots and then it is rapidly converted to nitrite (NO_2^-) by nitrate reductase (NR) (EC 1.6.6.1) which is a critical and very important enzyme for this conversion. This key reaction is mediated by transfer of two electrons from NAD(P)H to NO_3^-. This occurs via the three redox centers comprised of two prosthetic groups (FAD and heme) and a MoCo cofactor. NR activity is greatly influenced by factors such as developmental stage and various stress conditions such as hypoxia, salinity and pathogen infection etc. In addition, light/dark dynamics plays crucial role in modulating NR activity. NR activity can be easily detected by measuring the conversion of NO_3^- to NO_2^- under optimized conditions. Here, we describe a detailed protocol for measuring relative NR enzyme activity of tomato crude extracts. This protocol offers an efficient and straightforward procedure to compare the NR activity of various plants under different conditions.

Key words Nitrate reductase, Hypoxia, Light, Nitrate, Nitrite, Nitric oxide, Ammonium

1 Introduction

Nitrogen (N) is a crucial macronutrient required for plant, growth, development and conferring resistance against biotic and abiotic stresses [1]. In addition to these functions, N plays an important role in sensing and signaling processes. The Inorganic nitrogen assimilation pathway involves various important enzymes and cofactors. Among these enzymes cytosolic NR is one of the well-known and fully characterized enzymes. NR is a dimer that contains two identical subunits of the size 100–115 kDa [2]. NR

Kapuganti Jagadis Gupta (ed.), *Nitrogen Metabolism in Plants: Methods and Protocols*, Methods in Molecular Biology, vol. 2057, https://doi.org/10.1007/978-1-4939-9790-9_3,

(EC 1.6.6.1) catalyzes the first reaction in nitrate assimilation via the NAD(P)H dependent reduction of NO_3^- to NO_2^- in cytosol of cell. This is considered as a rate-limiting step [3].

$$NO_3^- + NAD(P)H + H^+ \rightarrow NO_2^- + H_2O + NAD(P)^+$$

Once NO_2^- is produced, it is transported into the plastids by putative nitrite transporters and reduced to NH_4^+ by nitrite reductase (NiR; EC 1.7.7.1) [4].

$$NO_2^- + 6_{Fdred} + 8H^+ + 6e^- \rightarrow NH_4^+ + 6_{Fdox} + 2H_2O$$

The produced NH_4^+ is then incorporated into carbon skeleton through glutamine synthetase and glutamate synthase (GS/GOGAT cycle) which is crucial cycle for amino acid biosynthesis. Operation of this process is tightly regulated and responsive to various factors such as FAD, heme b_{557}, and molybdenum cofactor (MoCo).

NR is regulated by both transcriptional and posttranslational levels [5, 6]. NO_3^- concentration differentially regulates its expression [7]. In Arabidopsis it was shown that NO_3^- and NH_4^+ treatment can differentially effect *NIA1* and *NIA2* expression [8]. Hypoxia and production of nitric oxide (NO) can also greatly influence *NIA1* and *NIA2* expression [9]. It was demonstrated that *NIA1* gene in highly induced under hypoxia but *NIA2* gene expression has decreased [9]. In addition to hypoxia, light can also affect its expression. Light and dark transition affects expression of nitrate reductase gene in maize [5]. For instance, when dark adapted etiolated seedlings or plants are transferred to light, then it induces the gene via phytochrome mediated signaling [10]. NODULE-INCEPTION-like proteins (NLPs) are transcription factors can modulate nitrate reductase encoded genes and also nitrate assimilation [11]. LONG HYPOCOTYL 5 (HY5) and HYH (HY5 Homologue) are bZIP transcription factors shows *NIR2* expression under continuous light conditions [12]. NR activity also depends on various factors such as light, pH, hypoxia ,and pathogen infection [13]. Light mediates post translational regulation of NR [6, 14]. However, an increased NR activity and opposing effect under dark period was observed [13, 15]. Accumulating evidence suggest that protein synthesis/degradation and phosphorylation/dephosphorylation are the key mechanisms for the regulation of NR activity during light–dark transition [16]. Under low cytosolic pH, NO_2^- reduction to NH_4^+ is greatly inhibited that leads to increase in production of gaseous free radical NO. This condition is most prevalent during hypoxia due to increased fermentation. It has been shown that NO_3^- and NO_2^- are major players in determining NR activity [13]. Furthermore, NO production from NR requires low nitrate and high nitrite [17]. NR dependent NO plays a role in stomatal moments during pathogen infection [18], drought tolerance via induction of several antioxidants [19], operation of phytoglobin nitric oxide (Pgb-NO) cycle for survival under hypoxia [9], aerenchyma formation [20], and hypersensitive response during avirulent pathogen infection to plants [21].

Here, we describe a detailed protocol for measuring NR activity by its final product, nitrite through its absorbance under three different conditions of Hypoxia/Normoxia, Light/Dark, and Pathogen/$MgCl_2$.

2 Materials

2.1 Nitrate Reductase Activity

2.1.1 Reagents

1. Extraction Buffer: 100 mM HEPES–KOH pH 7.5, 3.5 mM (*see* **Note 1**) mercaptoethanol, 0.3% Triton X-100, 15 mM $MgCl_2$, 0.5% PVP, 0.5% BSA.
2. Assay Buffer: 50 mM HEPES–KOH pH 7.5, 5 mM KNO_3, 0.2 mM NADH.
3. NED Solution: 0.1% *N*-1-napthylethylenediamine dihydrochloride (*see* **Note 2**).
4. Sulfanilamide Solution: 1% sulfanilamide in 5% phosphoric acid.
5. Stop solution: 0.5 M zinc acetate.

2.1.2 Strain Maintenance

1. *Pseudomonas syringae* pv. Tomato isolate DC3000.
2. King's medium B Base Agar (for 1 L): 20 g Proteose peptone, 1.5 g $MgSO_4$, 1.5 g K_2HPO_4, pH 7.0.
3. 10 mM $MgCl_2$.
4. Rifampicin Antibiotic (50 mg/l) (*see* **Note 3**).

2.2 Equipment

1. Centrifuge/centrifuge tubes.
2. Microplate Reader/Spectrophotometer.
3. Magnetic stirrer.
4. Stirring bar.
5. pH meter.
6. Autoclave.
7. Phytotron.
8. Hypoxia Chamber.
9. Mortar and pestle.
10. Spectrophotometer.

3 Methods

3.1 Plant Growth and Conditions

1. Sow the tomato seeds cv. Pusa Ruby (PR) in plastic trays filled with vermiculite-agropeat mixture in 3:1 ratio and moister the soil with sterile distilled water.
2. Germinate the seeds in phytotron set at 26 °C for a week under dark conditions.
3. After 7 days of germination, transfer the germinated seedlings to individual pots filled with sterile vermiculite–agropeat mixture.
4. Transfer the individual planted pots in phytotron which is set at a photoperiod of 16/8 h light/dark cycle, day/night temperature regime of 26°/20 °C, RH- 60% and photon flux density of 300 μmol m^{-2} s^{-1}.
5. Irrigate plants with water and half strength nutrient solution twice a week alternatively for a period of 7 weeks.

3.2 Treatments

3.2.1 Hypoxia Treatment

1. Detach tomato leaves from 7-week-old plants using scissors and stainless steel tweezers.
2. Dip petioles into the test tubes filled with 10 ml of 3 mM nitrate in 25 mM HEPES-KOH pH 7.5.
3. Transfer the test tube sets with leaves to hypoxic chamber for 2 h (Fig. 1). Cover the hypoxic chamber with black cloth to avoid photosynthesis.
4. Flash flush nitrogen gas for every 10 min until 2 h of complete treatment.

3.2.2 Light/Dark Treatment

1. Detach tomato leaves from 7-week -old plants described as above.
2. Prepare the test tubes with detached leaves as described earlier.
3. Incubate leaves in the phytotron with continuous light (300 μmol m^{-2} s^{-1}) for 2 h.
4. For dark treatment transfer leaves into dark room for 2 h.

3.2.3 Pathogen Preparation and Infiltration

1. Prepare 3 to 5 ml primary culture of *Pseudomonas syringae* pv. tomato isolate DC3000 strain in a 50 ml sterile glass culture vial containing sterile KB broth with rifampicin (50 mg/l) by adding an inoculum of 20 μl from its stored glycerol stock.
2. Incubate the primary culture at 28 °C shaker and continuously shake the culture at 220 rpm for a period of 24 h.
3. After incubation take few 10 μl of primary culture and add in to 100 ml of sterile KB broth with rifampicin (50 mg/l) in a sterile 250 ml conical flask for preparation of secondary culture (*see* **Note 4**).

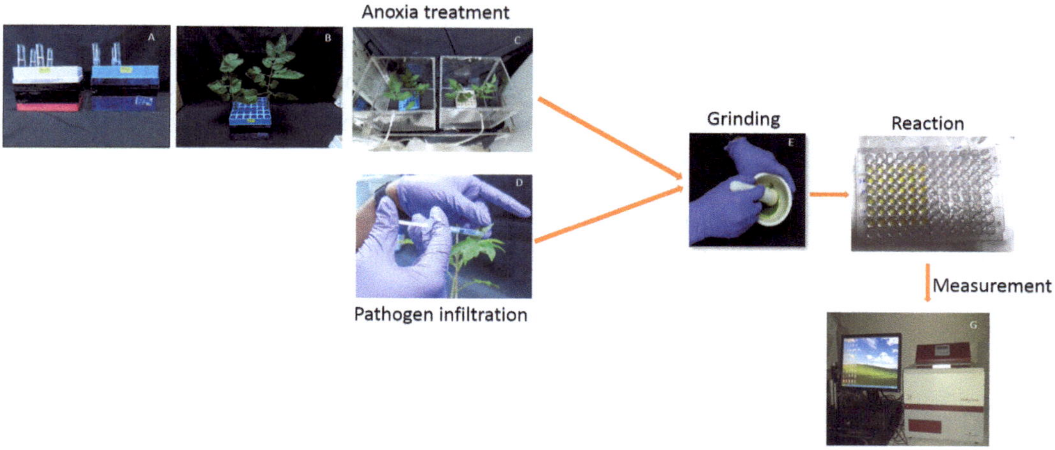

Fig. 1 Measurement of nitrite from anoxia (upper panel) and pathogen infected (lower panel) For anoxia treatment (**a**) Fill 5 mM nitrate solution (in 25 mM HEPES buffer, pH 7.5) filled in test tubes; (**b**) The petiole of the detached leaves dipped inside the buffer solution (**c**) Anoxia treatment to the leaves inside the customized hypoxia chamber (**d**) *PstDC3000* infiltration to the abaxial side of tomato leaves using syringe (**e**) Grinding of leaf tissue in mortar and pestle (**f**) Griess reaction set up in a 96-well microplate and (**g**) measurement of nitrite at 546 nm in a microplate reader

4. Incubate the secondary culture flask in shaker at 28 °C and shake 220 rpm for 12–16 h.
5. After incubation, transfer the secondary culture equally into two 50 ml sterile centrifuge tubes.
6. Centrifuge these tubes at 2100 × *g* for 10 min at room temperature.
7. Discard the supernatant in an autoclavable bag and resuspend pellet in 1 ml of sterile 10 mM $MgCl_2$ (*see* **Note 5**).
8. Measure the OD of the bacterial suspension at 600 nm in a spectrophotometer (*see* **Note 6**).
9. For tomato leaf infiltration use OD_{600} 0.2.
10. Transfer 7-week-old tomato plants from phytotron to the infection room.
11. Infiltrate the bacterial suspension ($OD_{600} = 0.2$) slowly on the abaxial surface of marked leaves using a 1 ml needleless syringe (*see* **Note 7**) (Fig. 1).
12. Allow the infiltrated leaves to dry and cover plants with a plastic wrap to maintain humidity.
13. Keep the infected plants in the infection room for 6 h, or as per required time (*see* **Note 8**).
14. For control treatment, similarly infiltrate other tomato plants as well with 10 mM $MgCl_2$.

3.3 Nitrate Reductase Assay

1. Harvest 0.5 g of tissue (FW) at different time points with a sterile stainless steel forceps and a scissors. Alternately scalpel blade can be used for harvesting tissue.
2. Label the different samples and immediately freeze in liquid nitrogen.

3.3.1 Preparation of Extract

1. Immediately after treatment grind preweighed tomato leaf tissue with liquid nitrogen into a fine powder using mortar and pestle (Fig. 1).
2. Homogenize the tissue powder in 500 μl of NR extraction buffer (Subheading 2.1.1) (*see* **Note 1**).
3. Centrifuge the extract at 16,800 × *g* (maximum speed) for 20 to 30 min at 4 °C to yield a clear supernatant.
4. The obtained aliquots of this supernatant can be kept on ice and directly used for the measurement of NR activity immediately.
5. Alternatively supernatant can be stored at −20° or −80 ° C for several months until further assay.

3.3.2 Griess Reaction for Measurement of Nitrite

1. Add 100–200 μl of the supernatant prepared at Subheading 3.3.1 to the 800 μl NR assay buffer (Subheading 2.1.1). After 5 min of incubation at 24 °C, stop the reaction by adding 125 μl of zinc acetate (0.5 M).
2. Transfer 100–200 μl of the supernatant into another centrifuge tube, then add 5 mM of KNO_3 and 0.2 mM of freshly prepared NADH with NR assay buffer and incubate the mixture for 10 min.
3. After incubation immediately stop this reaction by adding 100 μl zinc acetate (0.5 M) and add Griess reagent (0.5 ml sulfanilamide and 0.5 ml NED solution) and incubate for 10 min in dark.
4. Centrifuge at maximum speed for 2 to 5 min. Transfer 100 μl of solution from each treatment into the prelabeled 96-well microplate.
5. Measure the color (magenta) produced due to presence of nitrite using microplate reader set at 546 nm (Fig. 2) (*see* **Note 9**).
6. Calibrate the unknown values of nitrite with the precise nitrite standard reference curve to know the actual concentration of the NR extract in tomato leaves.
7. NR activity is expressed in nmoles/gFW/min or nmoles/mg protein/min (Fig. 3).

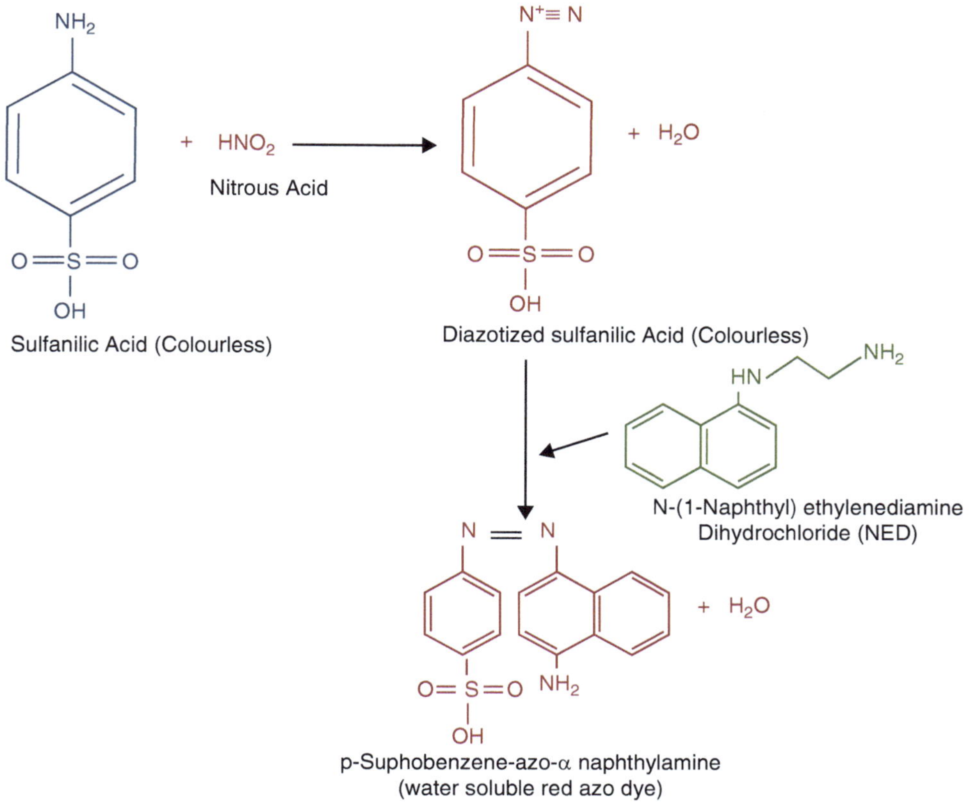

Fig. 2 Griess reagent reaction for measurement of nitrite. Nitrite first reacts with sulfanilic acid and forms a colorless diazotized sulfanilic acid which is then reacts with *N*-(1-naphthyl) ethylenediamine (NED). This overall reaction leads to formation of magenta colored sulfobenzene-azo-α-naphthylamine compound

4 Notes

1. Extraction buffer should be prepared fresh, used immediately and stored on ice for immediate activity measurement or stored at −20 °C or −80 °C for long term use.
2. *N*-(1-naphthyl) ethylenediamine solution should be stored in a dark bottle and if stored properly it can be stable for 1 month.
3. Rifampicin antibiotic is light sensitive, prepare stock in amber vials.
4. The incubation time and temperature, sterile condition of the primary and secondary culture should be followed strictly.
5. The bacterial pellet should always be resuspended in either sterile 10 mM $MgCl_2$ or $MgSO_4$ and should be done in laminar air flow.

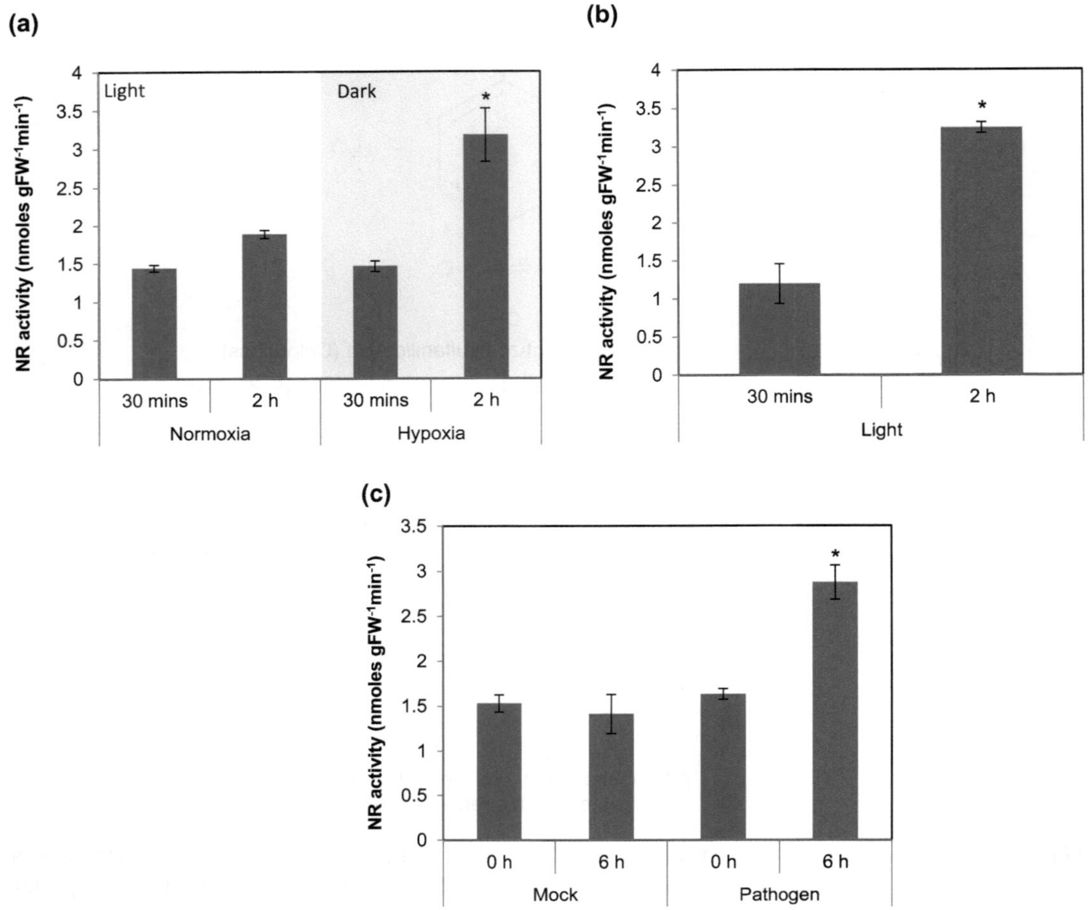

Fig. 3 Nitrate reductase (NR) activity in crude leaves extract of tomato (**a**) NR activity under normoxia and hypoxia dark conditions (30 min and 2 h) (**b**) NR activity in in response to 30 min and 2 h light treatment (**c**) NR activity in mock and Pst treated samples at 0 and 6 h post infiltration. Bars represent the means $\pm$ SE ($n = 3$). A significant difference within the treatments is denoted by $*$ at $p < 0.05$ (two sample t-test)

6. While taking OD at 600 nm, the proper handling of bacterial culture should be done. The cuvette can be shaked if required and the OD should be 0.2.
7. Bacterial or mock inoculation to abaxial side of the leaf should be done without imposing much pressure in order to avoid wounding effect.
8. The infection room should be different from the culture room and sterile conditions should be maintained and after experiment surface sterilization should be performed in the room.
9. Always perform the color development of the test and standards at the same time.

References

1. Gojon A (2017) Nitrogen nutrition in plants: rapid progress and new challenges. J Exp Bot 68:2457–2462
2. Solomonson LP, Barber MJ (1990) Assimilatory nitrate reductase: functional properties and regulation. Annu Rev Plant Biol 41:225–253
3. Campbell WH (2001) Structure and function of eukaryotic NAD (P) H: nitrate reductase. Cell Mol Life Sci 58:194–204
4. Stitt M (1999) Nitrate regulation of metabolism and growth. Curr Opin Plant Biol 2:178–186
5. Bowsher CG, Long DM, Oaks A, Rothstein SJ (1991) Effect of light/dark cycles on expression of nitrate assimilatory genes in maize shoots and roots. Plant Physiol 95:281–285
6. Kaiser WM, Spill D (1991) Rapid modulation of spinach leaf nitrate reductase by photosynthesis: II. In vitro modulation by ATP and AMP. Plant Physiol 96:368–375
7. Crawford NM, Smith M, Bellissimo D, Davis RW (1988) Sequence and nitrate regulation of the Arabidopsis thaliana mRNA encoding nitrate reductase, a metalloflavoprotein with three functional domains. Proc Natl Acad Sci USA 85:5006–5010
8. Wany A, Gupta AK, Aprajita Kumari, Mishra S, Singh N, Pandey S, Vanvari R, Igamberdiev AU, Fernie AR, Gupta KJ (2018) Nitrate nutrition influences multiple factors in order to increase energy efficiency under hypoxia in *Arabidopsis*. Ann Bot (In Press)
9. Vishwakarma A, Kumari A, Mur LAJ, Gupta KJ (2018) A discrete role for alternative oxidase under hypoxia to increase nitric oxide and drive energy production. Free Radic Biol Med. https://doi.org/10.1016/j.freeradbiomed.2018.03.045
10. Pilgrim ML, Caspar T, Quail PH, McClung CR (1993) Circadian and light-regulated expression of nitrate reductase in *Arabidopsis*. Plant Mol Biol 23:349–364
11. Castaings L, Camargo A, Pocholle D, Gaudon V, Texier Y, Boutet-Mercey S, Taconnat L, Renou JP, Daniel-Vedele F, Fernandez E, Meyer C (2009) The nodule inception-like protein 7 modulates nitrate sensing and metabolism in Arabidopsis. Plant J 57:426–435
12. Lee J, He K, Stolc V, Lee H, Figueroa P, Gao Y, Tongprasit W, Zhao H, Lee I, Deng XW (2007) Analysis of transcription factor HY5 genomic binding sites revealed its hierarchical role in light regulation of development. Plant Cell 19:731–749
13. Planchet E, Jagadis Gupta K, Sonoda M, Kaiser WM (2005) Nitric oxide emission from tobacco leaves and cell suspensions: rate limiting factors and evidence for the involvement of mitochondrial electron transport. Plant J 41:732–743
14. Kaiser WM, Huber SC (2001) Post-translational regulation of nitrate reductase: mechanism, physiological relevance and environmental triggers. J Exp Bot 52:1981–1989
15. Lillo C (1994) Light regulation of nitrate reductase in green leaves of higher plants. Physiol Plant 90:616–620
16. Heidari B, Matre P, Nemi-Feyissa D, Meyer C, Rognli OA, Moller SG, Lillo C (2011) Protein phosphatase 2A B55 and a regulatory subunits interact with nitrate reductase and are essential for nitrate reductase activation. Plant Physiol 156:165–172
17. Rockel P, Strube F, Rockel A, Wildt J, Kaiser WM (2002) Regulation of nitric oxide (NO) production by plant nitrate reductase in vivo and in vitro. J Exp Bot 53:103–110
18. Hao F, Zhao S, Dong H, Zhang H, Sun L, Miao C (2010) Nia1 and Nia2 are involved in exogenous salicylic acid-induced nitric oxide generation and stomatal closure in *Arabidopsis*. J Integr Plant Biol 52:298–307
19. Sang J, Jiang M, Lin F, Xu S, Zhang A, Tan M (2008) Nitric oxide reduces hydrogen peroxide accumulation involved in water stress-induced subcellular anti-oxidant defense in maize plants. J Integr Plant Biol 50:231–243
20. Wany A, Kumari A, Gupta KJ (2017) Nitric oxide is essential for the development of aerenchyma in wheat roots under hypoxic stress. Plant Cell Environ 40:3002–3017
21. Gupta M, Yoshioka H, Ohnishi K, Mizumoto H, Hikichi Y, Kiba A (2013) A translationally controlled tumor protein negatively regulates the hypersensitive response in *Nicotianabenthamiana*. Plant Cell Physiol 54:1403–1414

Chapter 4

Fluorimetric-Based Method to Detect and Quantify Total *S*-Nitrosothiols (SNOs) in Plant Samples

Paulo T. Mioto, Alejandra Matiz, Luciano Freschi, and Francisco J. Corpas

Abstract

Accumulating experimental evidence indicates that *S*-nitrosylation (technically *S*-nitrosation) events have a central role in plant biology, presumably accounting for much of the widespread influence of nitric oxide (NO) on developmental, metabolic, and stress-related plant responses. Therefore, the accurate detection and quantification of *S*-nitrosylated proteins and peptides can be particularly useful to determine the relevance of this class of compounds in the ever-increasing number of NO-dependent signaling events described in plant systems. Up to now, the quantification of *S*-nitrosothiols (SNOs) in plant samples has mostly relied on the Saville reaction and the ozone-based chemiluminescence method, which lacks sensitivity and are very time-consuming, respectively. Taking advantage of the photolytic properties of *S*-nitrosylated proteins and peptides, the method described in this chapter allows simple, fast, and high-throughput detection of SNOs in plant samples.

Key words Ozone chemiluminescence, Saville, *S*-nitrosothiols, *S*-nitrosoglutathione, Nitric oxide, Nitrosylated proteins

1 Introduction

Nitric oxide (NO) is a reactive nitrogen species (RNS) involved in a wide array of biological processes. In plants, it has been shown that RNS participate in abiotic and biotic stress responses and other key physiological processes, such as seed germination [1], root development [2, 3], leaf senescence [4, 5], stomatal closure [6–8], sexual reproduction [9–11], flowering [12], and fruit ripening [13, 14]. The capacity of NO to act through post-translational modification of proteins (PTMs) by *S*-nitrosylation (formally *S*-nitrosation) is one of the most important mechanisms that diversifies NO-related signals and may also act as a dynamic reservoir of NO [15–20]. *S*-nitrosylation occurs when NO reacts with the sulfhydryl group (–SH) of cysteine residues of target proteins or peptides, giving rise to *S*-nitrosothiols (SNOs). So far, there is an increasing amount of knowledge regarding which proteins can be

Kapuganti Jagadis Gupta (ed.), *Nitrogen Metabolism in Plants: Methods and Protocols*, Methods in Molecular Biology, vol. 2057, https://doi.org/10.1007/978-1-4939-9790-9_4, © Springer Science+Business Media, LLC, part of Springer Nature 2020

S-nitrosylated in plant cells, as well as the effects of this PTM on their function, demonstrating the relevance of this process under physiological and stress conditions [21–23]. In this context, the enzyme *S*-nitrosoglutathione (GSNO) reductase is a key element since it regulates the level of GSNO and consequently the level of SNOs [24–29].

Consequently, due to the importance of SNOs in plant responses and signaling mechanisms, it is essential to have several alternative and robust methods to detect and quantify these compounds [28, 30–34]. Nowadays, two main quantitative methods have been employed to analyze SNOs in plants: (1) The Saville reaction, which is based on the replacement of a nitroso (−NO) group from SNOs by mercury and subsequent formation of nitrite (NO_2^-). In this method, the samples are incubated in mercuric chloride ($HgCl_2$) and the formed nitrite is spectrophotometrically measured. The difference between absorbance with and without $HgCl_2$ indicates the levels of total SNOs (proteins and peptides) in the sample [30]. (2) The ozone chemiluminescence method, which is based on the release of NO upon reductive decomposition of nitroso species (−S−NO) from SNOs, with subsequent quantification of NO by using a gas-phase chemiluminescence [33]. The Saville reaction is one of the easiest and cheapest ways to quantify SNOs, but it lacks sensitivity [32, 34, 35]. In contrast, the ozone chemiluminescence-based method is highly sensitive, but it is time-consuming and requires sophisticate and rather expensive reaction and detection systems [32–34]. The present fluorescence-based method is simple and sensitive allowing the quantification of NO derived from SNOs by photolysis of the S-NO bond through UV light exposure. The emitted NO is detected spectrofluorometrically by using the NO-detecting fluorescent dye diaminorhodamine-4M (DAR-4M). The DAR-4M method proved to be more sensitive than the Saville assay, but not as sensitive as the ozone chemiluminescence. However, it is easy and relatively cheap to perform. Moreover, differently from the Saville and the ozone chemiluminescence-based assay, the DAR-4M method is not influenced by nitrite, a compound that may be found in relatively high concentrations in plant tissues.

2 Materials

2.1 Equipment

1. Measuring material (scales, pipettes, and other).
2. Vortex.
3. Refrigerated centrifuge (Sorvall™ ST 16R).
4. PCR plates (transparent 96-well 0.2 mL).

5. Transilluminator commonly used for visualizing the stained nucleic acids (Loccus L-PIX Touch).
6. Spectrofluorometer cuvettes (1 mL capacity).
7. Spectrofluorometer (Perkin-Elmer LS-55).

2.2 Reagents and Solutions

1. *S*-Nitrosoglutathione (GSNO–Sigma-Aldrich) 1 mM for standard curve–prepared fresh.
2. Diaminorhodamine-4M (DAR-4M–Calbiochem) 5 mM.
3. Chilled acetone—Keep in the freezer at least 1 h before the start of the protocol.
4. Phosphate buffer 50 mM, pH 7.2 with 80 mM of *S*-methyl methanethiosulfonate (MMTS). The buffer without MMTS can be stored at 4 °C, but the MMTS should be added on the same day of the analyses.
5. Phosphate buffer 50 mM, pH 7.2 (without MMTS—can be stored at 4 °C).

3 Methods

The fluorescence-based method uses the principle of cleavage of S-NO bonds by UV light. It was based on a procedure developed for animal samples [31], and a series of modifications were incorporated to ensure sensitive and reproducible measurements of plant samples of diverse origins [34]. Figure 1 shows a schematic workflow of the method.

1. During the whole procedure, keep the extracts on ice (4 °C) and in darkness whenever possible. Dim the lights in the room during the procedure, because SNOs usually are photosensitive.
2. Transfer the samples to the liquid nitrogen and grind into fine powder. Do not let it thaw.
3. Weight approximately 0.1 g of powdered tissue and add 1 mL of the phosphate buffer + MMTS (*see* **Note 1**). Vortex for 1 min. Centrifuge at 20,000 × *g* for 20 min at 4 °C.
4. Collect 0.5 mL of the supernatant and transfer to a 2 mL microtube. Add 1.5 mL of chilled acetone.
5. Leave for 20 min at −20 °C to precipitate proteins and peptides.
6. Centrifuge for another 20 min (20,000 × *g*; 4 °C).
7. Discard the supernatant and resuspend the pellet in 250 μL of the phosphate buffer + MMTS. Afterward, centrifuge for another 3 min to separate any insoluble material.

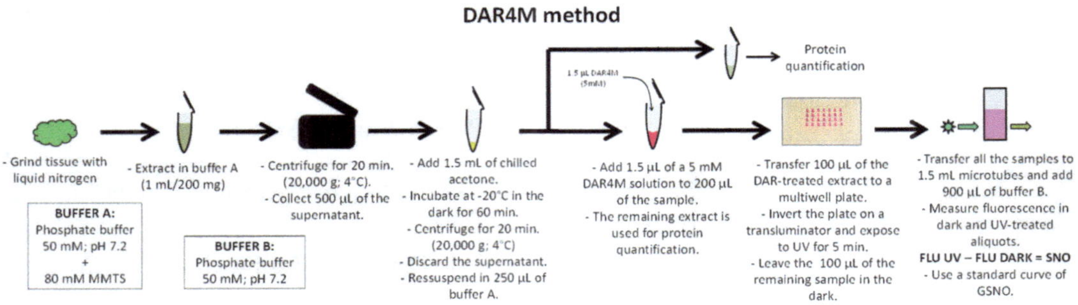

Fig. 1 Schematic workflow of the DAR-4M method

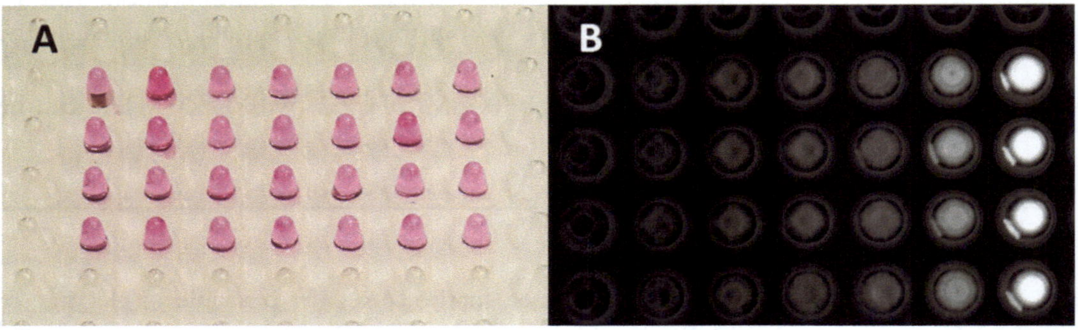

Fig. 2 (**a**) A 200 μL PCR plate inverted with liquid. (**b**) Visualization of the fluorescence of a GSNO standard curve during **step 11** of the protocol

8. Collect 200 μL of the supernatant into a new 1.5 mL microtube. Keep the remaining supernatant to determine protein concentration by the most convenient method—e.g., Bradford method [36].
9. Add 1.5 μL of the 5 mM DAR-4M solution to the 200 μL aliquot of your supernatant, to achieve a final concentration of 37.5 μM of the dye (*see* **Note 2**). Mix it well with the pipette. Do not vortex, nor shake the microtube.
10. Collect 100 μL of the supernatant + DAR-4M mix and accommodate it in a well of the PCR plate (*see* **Note 3**). Leave the remaining ~100 μL in total darkness at room temperature until **step 13**. This is the blank for your sample. Each sample must have a blank (Fig. 1).
11. Put the plate face down in a transilluminator for 5 min. The liquid will not fall unless the plate is vigorously shaken or hit (Fig. 2a). During this step, the fluorescence of the sample frequently can be seen, and the intensity roughly correlates with the SNOs content (Fig. 2b). Fluorescence should increase with time, so the time of exposure should be keep constant between different samples.

12. Carefully collect the entire sample from each well and transfer to 1.5 mL microtube. Add 900 μL of phosphate buffer (without MMTS) to each microtube. Also, add another 900 μL of phosphate buffer to the blank samples (the ~100 μL that remained in the tube during **step 10**).
13. Read all the blanks and samples in a spectrofluorometer with 560 nm excitation and 575 nm emission wavelength (5 nm bandwidth). All samples not currently being read must be kept in darkness, especially your blanks for each sample.
14. The difference in the fluorescence emission between each sample and its respective blank must be compared with a standard curve of GSNO to estimate the amount of SNOs.

4 Notes

1. The proportion between sample amount and extractions buffer may vary depending on species or treatment. Preliminary tests are necessary to determine the optimal proportion.
2. DAR-4M is photosensitive. Keep it protected from light whenever possible.
3. Carefully add the liquid into the well to avoid the formation of air bubbles. Use reverse pipetting if necessary.

Acknowledgments

Funding: This work was supported by Fundação de Amparo à Pesquisa do Estado de São Paulo (FAPESP) [grant numbers 2011/50637-0, 2013/18056-2 and 2013/15108-1], Conselho Nacional de Desenvolvimento Científico e Tecnológico (CNPq) [grant numbers 442045/2014-0, and 309504/2014-7], Coordenação de Aperfeiçoamento de Pessoal de Nível Superior (CAPES) [grant number 99999.006262/2015-05] and CNCSIS/UEFISCDI [grant PN-II-RU-TE-2014-4-2555]. Research in FJC laboratory is supported by the ERDF-cofinanced grants from the Ministry of Economy and Competitiveness (AGL2015-65104-P) and Junta de Andalucía (group BIO192), Spain.

References

1. Bethke PC, Libourel IG, Jones RL (2006) Nitric oxide reduces seed dormancy in Arabidopsis. J Exp Bot 57:517–526
2. Pagnussat GC, Lanteri ML, Lamattina L (2003) Nitric oxide and cyclic GMP are messengers in the indole acetic acid-induced adventitious rooting process. Plant Physiol 132:1241–1248
3. Correa-Aragunde N, Graziano M, Chevalier C, Lamattina L (2006) Nitric oxide modulates the expression of cell cycle regulatory genes during lateral root formation in tomato. J Exp Bot 57:581–588
4. Corpas FJ, Barroso JB, Carreras A, Quirós M, León AM, Romero-Puertas MC, Esteban FJ, Valderrama R, Palma JM, Sandalio LM,

Gómez M, del Río LA (2004) Cellular and subcellular localization of endogenous nitric oxide in young and senescent pea plants. Plant Physiol 136:2722–2733

5. Wang Y, Loake GJ, Chu C (2013) Cross-talk of nitric oxide and reactive oxygen species in plant programed cell death. Front Plant Sci 4:314
6. Bright J, Desikan R, Hancock JT, Weir IS, Neill SJ (2006) ABA-induced NO generation and stomatal closure in Arabidopsis are dependent on H_2O_2 synthesis. Plant J 45:113–122
7. Neill S, Barros R, Bright J, Desikan R, Hancock J, Harrison J, Morris P, Ribeiro D, Wilson I (2008) Nitric oxide, stomatal closure, and abiotic stress. J Exp Bot 59:165–176
8. Agurla S, Gayatri G, Raghavendra AS (2014) Nitric oxide as a secondary messenger during stomatal closure as a part of plant immunity response against pathogens. Nitric Oxide 43C:89–96
9. Prado AM, Colac OR, Moreno N et al (2008) Targeting of pollen tubes to ovules is dependent on nitric oxide (NO) signaling. Mol Plant 1:703–714
10. Bright J, Hiscocl SJ, James PE et al (2009) Pollen generates nitric oxide and nitrite: a possible link to pollen-induced allergic responses. Plant Physiol Biochem 47:49–55
11. Jiménez-Quesada MJ, Carmona R, Lima-Cabello E, Traverso JA, Castro AJ, Claros MG, Alché JD (2017) Generation of nitric oxide by olive (Olea europaea L.) pollen during in vitro germination and assessment of the S-nitroso- and nitro-proteomes by computational predictive methods. Nitric Oxide 68:23–37
12. Simpson GG (2005) NO flowering. BioEssays 27:239–241
13. Manjunatha G, Lokesh V, Neelwarne B (2010) Nitric oxide in fruit ripening: trends and opportunities. Biotechnol Adv 28:489–499
14. Corpas FJ, Freschi L, Rodríguez-Ruiz M et al (2018) Nitro-oxidative metabolism during fruit ripening. J Exp Bot. https://doi.org/10.1093/jxb/erx453
15. Sokolovski S, Blatt MR (2004) Nitric oxide block of outward rectifying K^+ channels indicates direct control by protein nitrosylation in guard cells. Plant Physiol 136:4275–4284
16. Feechan A, Kwon E, Yun B-W et al (2005) A central role for S-nitrosothiols in plant disease resistance. Proc Natl Acad Sci U S A 102 (22):8054–8059
17. Tada Y, Spoel SH, Pajerowska-Mukhtar K et al (2008) Plant immunity requires conformational changes [corrected] of NPR1 via S-nitrosylation and thioredoxins. Science 321:952–956
18. Yun BW, Feechan A, Yin M et al (2011) S-nitrosylation of NADPH oxidase regulates cell death in plant immunity. Nature 478:264–268
19. Fares A, Rossignol M, Peltier JB (2011) Proteomics investigation of endogenous S-nitrosylation in Arabidopsis. Biochem Biophys Res Commun 416:331–336
20. Wang P, Du Y, Hou YJ et al (2015) Nitric oxide negatively regulates abscisic acid signaling in guard cells by S-nitrosylation of OST1. Proc Natl Acad Sci U S A 112:613–618
21. Astier J, Kulik A, Koen E, Besson-Bard A, Bourque S, Jeandroz S, Lamotte O, Wendehenne D (2012) Protein S-nitrosylation: what's going on in plants? Free Radic Biol Med 53:1101–1110
22. Begara-Morales JC, Sánchez-Calvo B, Chaki M, Valderrama R, Mata-Pérez C, Padilla MN, Corpas FJ, Barroso JB (2016) Antioxidant systems are regulated by nitric oxide-mediated post-translational modifications (NO-PTMs). Front Plant Sci 7:152
23. Zaffagnini M, De Mia M, Morisse S, Di Giacinto N, Marchand CH, Maes A, Lemaire SD, Trost P (2016) Protein S-nitrosylation in photosynthetic organisms: a comprehensive overview with future perspectives. Biochim Biophys Acta 1864:952–966
24. Chaki M, Valderrama R, Fernández-Ocaña AM, Carreras A, Gómez-Rodríguez MV, Pedrajas JR, Begara-Morales JC, Sánchez-Calvo B, Luque F, Leterrier M, Corpas FJ, Barroso JB (2011) Mechanical wounding induces a nitrosative stress by down-regulation of GSNO reductase and an increase in S-nitrosothiols in sunflower (Helianthus annuus) seedlings. J Exp Bot 62:1803–1813
25. Airaki M, Sánchez-Moreno L, Leterrier M, Barroso JB, Palma JM, Corpas FJ (2011) Detection and quantification of S-nitrosoglutathione (GSNO) in pepper (Capsicum annuum L.) plant organs by LC-ES/MS. Plant Cell Physiol 52:2006–2015
26. Corpas FJ, Alché JD, Barroso JB (2013) Current overview of S-nitrosoglutathione (GSNO) in higher plants. Front Plant Sci 4:126
27. Frungillo L, Skelly MJ, Loake GJ, Spoel SH, Salgado I (2014) S-nitrosothiols regulate nitric oxide production and storage in plants through the nitrogen assimilation pathway. Nat Commun 5:5401
28. Rodríguez-Ruiz M, Mioto PT, Palma JM et al (2017) Detection of protein S-nitrosothiols

(SNOs) in plant samples on diaminofluorescein (DAF) gels. Bio-Protocol 7(18):e2559
29. Lindermayr C (2017) Crosstalk between reactive oxygen species and nitric oxide in plants: key role of S-nitrosoglutathione reductase. Free Radic Biol Med. pii: S0891-5849(17) 31231-5
30. Saville B (1958) A scheme for the colorimetric determination of microgram amounts of thiols. Analyst 83:670–672
31. King M, Gildemeister O, Gaston B et al (2005) Assessment of S-nitrosothiols on diaminofluorescein gels. Anal Biochem 346:69–76
32. Diers AR, Keszler A, Hogg N (2014) Detection of S-nitrosothiols. BBA-Gen Subj 1840:892–900
33. Barroso JB, Valderrama R, Carreras A et al (2016) Quantification and localization of S-nitrosothiols (SNOs) in higher plants. Methods Mol Biol 1424:139–147
34. Mioto PT, Rodriguez-Ruiz M, Mot AC et al (2017) Alternative fluorimetric-based method to detect and compare total S-nitrosothiols in plants. Nitric Oxide 68:7–13
35. Palmer L, Gaston B (2008) S-nitrosothiol assays that avoid the use of iodine. Method Enzymol 440:157–176
36. Bradford MM (1976) A rapid and sensitive method for the quantitation of microgram quantities of protein utilizing the principle of protein-dye binding. Anal Biochem 72:248–254

Chapter 5

Measurement of *S*-Nitrosoglutathione Reductase Activity in Plants

Martina Janků, Tereza Tichá, Lenka Luhová, and Marek Petřivalský

Abstract

S-nitrosation as a redox-based posttranslational modification of protein cysteine has emerged as an integral part of signaling pathways of nitric oxide across all types of organisms. Protein *S*-nitrosation status is controlled by two key mechanisms: by direct denitrosation performed by the thioredoxin/thioredoxin reductase system, and in an indirect way mediated by *S*-nitrosoglutathione reductase (GSNOR). GSNOR, which has been identified as a key component of *S*-nitrosothiols catabolism, catalyzes an irreversible decomposition of abundant intracellular *S*-nitrosothiol, *S*-nitrosoglutathione (GSNO) to oxidized glutathione using reduced NADH cofactor. In plants, GSNOR has been shown to play important roles in plant growth and development and plant responses to abiotic and biotic stress stimuli. In this chapter, optimized protocols of spectrophotometric measurement of GSNOR enzymatic activity and activity staining in native polyacrylamide gels in plant GSNOR are presented.

Key words *S*-nitrosation, *S*-nitrosothiols, Nitric oxide, *S*-nitrosoglutathione reductase, Plant stress

1 Introduction

Nitric oxide (NO), a reactive gaseous compound, has been identified as an important signaling and regulatory molecule in diverse physiological and pathological processes in all types of organisms. Signaling pathways of NO are involved in the regulation of plant growth and development and growth and also in orchestration of plant responses to environmental stress factors [1]. NO-dependent signaling event are mediated mainly by fast NO reaction with reactive oxygen species (ROS), by nitrosylation of transition metal ions within protein structures and nitrosative modification of thiols groups in proteins and low-molecular weight thiols. Cysteine *S*-nitrosation has been recognized among important posttranslational protein modifications and a multiple *S*-nitrosation targets have been detected among plant proteins including enzymes, transcription factors, and structural proteins [2]. *S*-nitrosoglutathione (GSNO), as the most abundant low molecular weight *S*-

Kapuganti Jagadis Gupta (ed.), *Nitrogen Metabolism in Plants: Methods and Protocols*, Methods in Molecular Biology, vol. 2057, https://doi.org/10.1007/978-1-4939-9790-9_5,

nitrosothiol can under specific conditions release free NO radical or participate in protein transnitrosation, [3]. Similarly to animals, GSNO is considered as a stable form of NO bioactivity in plants in vivo, capable to transfer NO-dependent signals both in local and systemic responses of plant tissues [4]. Several reaction pathways of combined NO and ROS reactivity are known to contribute to the GSNO pool in plant cell; however, a specific enzymatic *S*-nitrosoglutathione reductase (GSNOR) seems to exert crucial regulatory role in GSNO decomposition and thus indirectly also in the regulation of intracellular levels of NO and protein *S*-nitrosothiols [5].

GSNOR activity was originally identified as a physiologically relevant activity of a previously characterized enzyme, *S*-(hydroxymethyl)glutathione dehydrogenase (EC 1.1.1.284, formerly known also as glutathione-dependent formaldehyde dehydrogenase, EC 1.2.1.1), which catalyzes the conversion of *S*-(hydroxymethyl)glutathione (HMGSH), a spontaneous adduct of glutathione and formaldehyde, to *S*-formylglutathione within the formaldehyde detoxification pathway [6]. However, the denomination of this enzyme as GSNOR has been widely accepted due to biological relevance of NADH-dependent reduction of GSNO to oxidized glutathione and ammonium [7, 8]. Based on its structural characteristics, GSNOR is classified among the family of Zn-dependent class III alcohol dehydrogenases (ADH3; EC 1.1.1.1), recognizable from the class I ADH family by their high affinity to long-chain alcohols [9]. Nevertheless, the comparison of kinetic parameters for its substrates suggest GSNO is the physiologically relevant substrate as the K_m value for HMGSH is approx. two-orders of magnitude higher compared to GSNO [10].

Using both pharmacological approaches and mutant plants with up-and down-regulated GSNOR, it has been confirmed that GSNOR act as a master regulator of protein *S*-nitrosylation during plant development and responses to abiotic and biotic stress stimuli [11]. Plant GSNORs are unusually cysteine-rich proteins with cytosolic localization, although also peroxisomal targeting has been suggested by some previous reports [12]. GSNOR activity can be determined as the amount of NADH consumed in the enzyme-dependent GSNO reduction by monitoring the absorbance decrease at 340 nm, whereas in the dehydrogenase mode, the oxidation of HMGSH is evaluated by the increase of absorbance at 340 nm. Native polyacrylamide gel electrophoresis (PAGE), which can achieve efficient separation of sample proteins under nondenaturing conditions by can be exploited to detect enzyme activities and analyze isoenzyme pattern in plant extracts. GSNOR reductase activity in the native PAGE gels can be followed by recording of decreased fluorescence resulting from the oxidation of NADH. Gels are sequentially incubated in filter papers soaked in buffers containing the enzyme cofactor NADH and then the enzyme substrate GSNO, where GSNOR activity is revealed as dark bands on gels to fluorescence-excitation illumination [13, 14]. Alternatively, GSNOR dehydrogenase activity can be stained after incubation with suitable substrates and electron acceptors such as nitroblue tetrazolium (NBT), which are reduced insoluble formazans observed as colored bands at the site of GSNOR presence in the gels. HMGSH, which can be prepared by mixing of corresponding solutions of reduced glutathione and formaldehyde, or long-chain alcohols like octanol are usually used as substrates of GSNOR/ADH3 dehydrogenase activity [15]. The dehydrogenase staining of GSNOR in native PAGE gels has the advantage of higher stability of used compounds and easier detection of visible bands compared to requirements and equipment needed for the detection of NADH fluorescence.

2 Materials

All solutions shall be prepared from chemicals of analytical grade using ultra-pure water or HPLC-grade solvents and stored at room temperature, unless indicated otherwise. Rules of safety disposal and regulations of waste material disposal shall be followed.

2.1 Synthesis of S-Nitrosoglutathione (GSNO)

1. GSNO is synthetized from the reduced glutathione (GSH) using acidified nitrite in HCl [16] (*see* **Note 1**).
2. Prepare 1 M HCl by adding drop-wise 4.3 mL of concentrated HCl (36%) to 40 mL of water and adjust final volume with water to 50 mL.
3. Dissolve 614 mg of reduced GSH (2 mM concentration) in 3 mL of 1 M HCl and place the solution on ice bath (*see* **Note 2**).
4. Add slowly 138 mg of sodium nitrite ($NaNO_2$, 2 mM) to the acidic glutathione solution under vigorous stirring on ice bath (*see* **Note 3**). The reaction mixture shall be stirred for at least 40 min on ice.
5. To precipitate synthetized GSNO out of the solution, add 5 mL of ice-cold acetone and stir for another 10 min on ice (*see* **Note 4**).
6. Separate the precipitated GSNO from the reaction mixture using a paper filter placed on a Buchner funnel and wash the precipitate with successive portions of ice-cold water (2 times 10 mL), ice-cold acetone (2 times 10 mL) and ice-cold diethyl ether (2 times 10 mL).
7. Dry the washed GSNO precipitate protected from the light (i.e., in an aluminum foil) in desiccator under vacuum for 30 min. Store the final GSNO product at −20 °C (*see* **Note 5**).

2.2 Sample Preparation

1. Extraction buffer: 50 mM Tris–HCl pH 7.5, 0.2% Triton X-100, 2 mM DTT, 1 mM (4–2-aminoethyl)benzenesulfonyl fluoride hydrochloride (AEBSF) (*see* **Note 6**). Weigh 0.606 g Tris and 0.2 g Triton X-100 and dissolve in 75 mL of water, mix and adjust pH with concentrated HCl. Make up to 100 mL with water and store at 4 °C. DTT and AEBSF should be added freshly before the use. To prepare 20 mL of the extraction buffer, dissolve 6.2 mg DTT and 4.8 mg AEBSF in approx. 15 mL of the TRIS-HCl buffer, than make up to 20 mL with water.
2. Equilibration and elution buffer: 20 mM Tris–HCl, pH 8.0. Weight 2.423 g Tris and dissolve in 950 mL of water, mix and adjust pH with HCl and make up to 1 L with water. Store at 4 °C.

2.3 Protein Quantification by Bradford Method

1. Stock solution of Coomassie Brilliant Blue (CBB): Weigh 50 mg CBB G250, dissolve it in 25 mL of methanol, add 50 mL of 85% H_3PO_4 and make up to 100 mL with water (*see* **Note 7**).
2. Protein standard: 10 mg/mL bovine serum albumin (BSA) in water. Aliquots can be stored at −20 °C.

2.4 Measurement of GSNOR Reductase Activity

1. Reaction buffer: 20 mM Tris–HCl, pH 8.0. Weigh 2.423 g Tris and dissolve in 950 mL of water, mix and adjust pH with HCl and make up to 1 L with water. Store at 4 °C.
2. Working solution of NADH (2 mM): Weigh 1.42 mg NADH disodium salt hydrate (Sigma-Aldrich, N8129) and dissolve in 1 mL of water. Should be prepared fresh on daily basis and kept on ice during the measurements.
3. Working solution of GSNO (4 mM) GSNO: Weigh 1.34 mg GSNO and dissolve in 1 mL of water. Should be prepared fresh on daily basis and kept on ice protected from the light (*see* **Note 8**).

2.5 Measurement of GSNOR Dehydrogenase Activity

1. Reaction buffer: 20 mM Tris–HCl, pH 8.0. Weigh 2.423 g Tris and dissolve in 950 mL of water, mix and adjust pH with concentrated HCl and make up to 1 L with water. Store at 4 °C.
2. Working solution of NAD^+ (60 mM): Weigh 40 mg NAD^+ hydrate (Sigma-Aldrich, N1636) and dissolve it in 1 mL of water. Should be prepared fresh on daily basis.
3. Glutathione solution (40 mM): Weigh 123 mg GSH and dissolve it in 10 mL of water.
4. Formaldehyde solution (31.2 mM): Dilute 31 μL of 30% formaldehyde (about 10 M) to 10 mL of water (*see* **Note 9**).

2.6 Detection of In-Gel GSNOR Activity

1. GSNOR reductase activity (fluorescence detection): weigh either 7 mg NADH or 6.5 mg GSNO and dissolve them in separate 5 mL volumes of 20 mM TRIS-HCl buffer, pH 8.0, respectively. Prior the use, keep both solutions of NADH and GSNO at 4 °C protected from the light.
2. GSNOR dehydrogenase activity (color staining): Enzyme activity is detected by gel incubation in 0.1 M sodium phosphate buffer (pH 7.4) containing 0.1 mM NAD^+, 0.1 mM nitroblue tetrazolium (NBT), 0.1 mM phenazine methosulfate (PMS), 1 mM reduced glutathione (GSH), and 1 mM formaldehyde. Weigh 6.6 mg NAD^+, 8.2 mg NBT, 3.1 mg PMS, and 30.7 mg GSH, dissolve weighed compounds sequentially in given order in 100 mL of 20 mM TRIS-HCl buffer, pH 8.0, and finally add 30 μL of formaldehyde. Store staining solution in dark prior to use (*see* **Note 10**).

2.7 Preparation of Native Polyacrylamide Gels

1. Resolving gel buffer: 1.5 M Tris–HCl, pH 8.8. Weigh 36.3 g Tris and dissolve in 150 mL of water. Adjust pH with HCl and make up to 200 mL with water. Store at 4 °C.
2. Stacking gel buffer: 0.5 M Tris–HCl, pH 6.8. Weigh 6.1 g Tris and dissolve in approx. 80 mL of water. Adjust pH with HCl and make up to 100 mL with water. Store at 4 °C.
3. Acrylamide–bisacrylamide (BIS) stock solution (30:0.8 acrylamide–BIS). Weigh 30 g acrylamide monomer and 0.8 g BIS (cross-linker) and transfer to a glass beaker. Add water to a volume of 60 mL and mix for about 30 min (*see* **Note 11**). Make up to 100 mL with water. Store at 4 °C protected from light.
4. Ammonium persulfate (APS, 10% solution in water): Weigh 25 mg APS and dissolve in 250 μL of water. Always prepare fresh just prior the use (*see* **Note 12**).
5. *N*, *N*, *N*′,*N*′-tetramethylethylenediamine (TEMED). Store at 4 °C.
6. 60% glycerol (v/v) in water.
7. 0.02% bromophenol blue in 20% glycerol (v/v).
8. Running buffer (25 mM Tris–HCl, 0.192 M glycine, pH 8.3): Weigh 6.05 g Tris, 28.82 g glycine (electrophoresis grade, e.g., Sigma-Aldrich G8898) and transfer to a glass beaker. Make up to 2 L with water, do not adjust pH. Store at 4 °C (*see* **Note 13**).

3 Methods

All procedures shall be performed at room temperature, unless otherwise specified.

3.1 Sample Extraction and Desalting

1. Add two volumes of extraction buffer (v/w) to the plant samples and homogenize e.g., using a mortar and pestle in liquid nitrogen.
2. Centrifuge crude extracts 20 min at 16,000 × *g* at 4 °C. Transfer carefully obtained supernatants into clean microtubes and store at 4 °C.
3. Equilibrate desalting column (PD-10 or NAP-10 columns, GE Healthcare, USA) with at least 15 mL of equilibration buffer.
4. Add 1 mL of sample extract to the column and let the sample to enter completely into the packed bed (*see* **Note 14**). Do not collect the column eluate at this step.
5. Add 2 mL of elution buffer to the column and collect the column eluate containing the protein fraction to a clean 2 mL microtube. Store the eluate at 4 °C (*see* **Note 15**).

3.2 Total Protein Assay by Bradford Method

1. Prepare a set of BSA solutions in the concentration range of 0.05–1.4 mg protein/mL by dilution of stock 10 mg/mL BSA solution (*see* **Note 16**).
2. Prepare working Bradford reagent by diluting the stock CBB G-250 solution with water in ratio 1:4 (*see* **Note 17**).
3. Pipet 5 μL of BSA protein standard, extract sample or blank to microplate wells. Elution buffer used in the extract desalting step should be used as a blank. Protein samples, standards and blanks are usually assayed at least in technical triplicates. Samples with the protein content higher than 1 mg/mL should be appropriately diluted with the elution buffer.
4. Add 200 μL of Bradford reagent to each microplate well, mix thoroughly and incubate 10 min at room temperature (*see* **Note 18**).
5. Measure absorbance values at 595 and 450 nm and calculate absorbance ratios as A_{595}/A_{450}. Subtract the mean of blank A_{595}/A_{450} ratio values from the measured A_{595}/A_{450} ratio values of standards and samples (*see* **Note 19**).
6. Protein sample concentrations are determined using standard curves obtained by a linear regression of A_{595}/A_{450} ratio values of protein standards versus their concentrations.

3.3 Measurement of GSNOR Reductase Activity

1. GSNOR reductase activity is assayed spectrophotometrically at 25 °C by monitoring oxidation of the enzyme cofactor NADH at 340 nm (*see* **Note 20**).
2. Prepare fresh 2 mM NADH and 4 mM GSNO. Store at 4 °C in dark.
3. For microplate assay, add 225 μL of Tris–HCl reaction buffer, 15 μL of the desalted sample and 30 μL of 2 mM NADH to 96-well microplates in triplicate.
4. The reaction is started by addition of 30 μL of 4 mM GSNO; as the blank use 30 μL of water (*see* **Note 21**).
5. Measure the decrease of absorbance at 340 nm during 1–10 min interval (*see* **Note 22**).
6. The GSNOR reductase activity is expressed as nmol of NADH consumed per sec and mg of protein, using extinction coefficient of NADH $\varepsilon = 6.22\ mM^{-1}\ cm^{-1}$ (*see* **Note 23**).

3.4 Measurement of GSNOR Dehydrogenase Activity

1. GSNOR dehydrogenase activity is assayed spectrophotometrically at 25 °C by monitoring the formation of the reduced cofactor NADH at 340 nm with the glutathione-formaldehyde adduct *S*-(hydromethyl)glutathione (HMGSH) as the enzyme substrate (*see* **Note 24**).
2. Prepare fresh solutions of 60 mM NAD^+, 40 mM reduced glutathione and 31.2 mM formaldehyde. Store all solutions at 4 °C.
3. In a 2 mL microtube, mix 1 mL of 40 mM GSH and 1 mL of 31.2 mM formaldehyde to obtain 2 mL of 20 mM HMGSH (*see* **Note 25**).
4. For the microplate assay, add 205 μL of 20 mM Tris–HCl reaction buffer, 15 μL of the desalted sample and 10 μL of 60 mM NAD^+ to 96-well microplate in triplicate.
5. The reaction is started by adding 15 μL of 20 mM HMGSH; as the blank use 15 μL of water instead of the substrate (*see* **Note 21**).
6. Measure the increase of absorbance at 340 nm during 1–10 min interval (*see* **Note 22**).
7. The GSNOR dehydrogenase activity is expressed as nmol NADH produced per sec and mg protein ($\varepsilon = 6.22\ mM^{-1}\ cm^{-1}$) (*see* **Note 23**).

3.5 Protein Separation by Native Polyacrylamide Gel Electrophoresis

1. The instructions described herein are specific to preparation of two 0.75 mm-thick minigels on the Mini-PROTEAN® Tetra System (Bio-Rad, USA) and can be correspondingly adjusted to prepare larger or thicker gels (*see* **Note 26**).
2. Clean thoroughly before every use both short and spacer plates cleaned with deionized water and finally with 95% ethanol.

3. Assemble two pairs of short and 0.75 mm spacer plate and insert them into the gel casting stand. Check tightness of the casting plate assembly by pouring deionized water between the plates. Remove the water and let the plates to completely dry.
4. Prepare the solution resolving gel solution by mixing 2.5 mL of resolving buffer, 3.3 mL of 30% acrylamide–BIS mixture and 4.1 mL of deionized water in a 50 mL conical flask (*see* **Note 27**).
5. Degas the solution for at least 10 min under vacuum and continuous stirring (*see* **Note 28**).
6. Add 10 μL of TEMED to the degassed resolving gel solution.
7. Add 100 μL of freshly prepared APS solution under continuous stirring and after that pour the gel solution immediately between assembled glass plates. The solution level should reach approximately 1 cm below the lower end of the comb (*see* **Note 29**).
8. Overlay the casted resolving gel solution with a thin layer of n-butanol saturated with water using a suitable pipette and allow to stand for approx. 30 min to enable gel polymerization (*see* **Note 30**).
9. Remove butanol from the resolving gel and rinse the gel surface with water. Remove the water and dry the area between the glass plates above the gel surface using a piece of filter paper.
10. Prepare the stacking gel solution by mixing 1.25 mL of resolving buffer, 0.65 mL of 30% acrylamide–BIS mixture, and 3.05 mL of water in a 50 mL conical flask (*see* **Note 27**).
11. Degas this solution for at least 10 min under vacuum and continuous stirring (*see* **Note 28**).
12. Add 10 μL of TEMED to the degassed gel solution.
13. Add 50 μL of APS solution under continuous stirring and pour immediately the solution on the top of stacking gel between assembled glass plates. The level of the solution should reach the top of the short plates.
14. Immediately insert a 0.75 mm comb into the space between the glass plates without introducing air bubbles and let the stacking gel to polymerize for approx. 30 min (*see* **Note 29**).
15. Prepare protein samples by mixing desalted plant extracts with 60% glycerol in ratio 3:1 (v/v) (*see* **Note 31**).
16. When the polymerization of stacking gels is finished, take out the glass plates with casted polyacrylamide gels from the casting stand, but keeping the comb inside the stacking gel.
17. Assemble the electrophoretic cell according the manufacturer instructions and insert the plates carrying the polyacrylamide gel facing the short plates inside the cell.

18. Fill assembled electrophoretic cell with prepared the electrode buffer to a level above the end of short plates, then carefully remove the combs from the stacking gels and rinse the wells in stacking gel with the electrode buffer thoroughly using a Pasteur pipette (*see* **Note 32**).
19. Check for air bubbles occasionally occurring at the bottom of the electrophoresis cell under the glass plates - if present, remove the bubbles using a Pasteur pipette (*see* **Note 33**).
20. Using a Hamilton syringe or gel-loader pipette tips, transfer 10–20 μL of prepared protein samples slowly into selected wells. Do not use the two outer wells for sample loading, instead load there a small amount of 0.02% bromophenol blue in 20% glycerol (up to 5 μL) to follow visually the progress of the electrophoretic run (*see* **Note 34**).
21. Place a lid on the electrophoretic cell, connect the cables to a suitable power supply and set the power supply to a value of 100 V to initiate the electrophoretic separation of sample proteins (*see* **Note 35**).
22. After the bromophenol blue zone reaches the upper limit of the separation gel, increase the voltage to 150 V. Continue the separation until the bromophenol blue migrates out from the gel. Then turn off the power supply and disconnect the cables from the power supply.
23. Disassemble the electrophoresis cell, pull out the glass plates with the gel and gently remove the gel by slowly separating the glass plates. Cut and discard the stacking gel and mark the upper part of the running gel by cutting one of its lower corners (*see* **Note 36**).
24. If required, separated proteins can be detected in polyacrylamide gels by any established method of total protein staining like CBB or silver staining.

3.6 Detection of GSNOR Activity in Gels

3.6.1 Fluorescence Detection of GSNO Reductase Activity

1. Prepare four 8 × 5 cm pieces of thick (~1 mm) filter paper (*see* **Note 37**).
2. After the end of electrophoretic separation, remove the stacking gels to a trash container and transfer with caution the separating gels to a suitable plastic tray (*see* **Note 38**).
3. Wash gels for 5 min in deionized water. Place evenly washed gels on the surface of UV tray for fluorescence imaging (*see* **Note 39**).
4. Soak two filter papers in 20 mM TRIS-HCl buffer, pH 8.0, containing 2 mM NADH (*see* **Note 40**).
5. Place soaked filter papers on the gel surface and remove the air bubbles between the gel and filter paper by rolling over a glass bar.

6. Incubate the gels in dark for 15 min and then remove the filter papers soaked with NADH.
7. Prepare two filter papers soaked with 20 mM TRIS-HCl buffer, pH 8.0, containing 4 mM GSNO (*see* **Note 40**).
8. Place soaked filter paper on the surface of the gel and eliminate the air bubbles between the gel and filter paper using a glass bar.
9. Incubate the gels in dark for 15 min and then remove the filter papers.
10. Place immediately the tray inside the documentation system (*see* **Note 41**).
11. Set the exposure parameters using suitable UV excitation and emission filter settings and start gel imaging (*see* **Note 42**).
12. Finally, the presence of GSNOR activity is associated with appearance of dark bands on the gel, corresponding to the disappearance of NADH fluorescence.
13. If required, a semiquantitative comparison of signal intensities of detected bands can be performed using available software tools for image analysis (proprietary software of documentation system or any available freeware as Image Lab).

3.6.2 Staining for GSNOR Dehydrogenase Activity

1. Prepare suitable plastic tray and place the separation gel inside.
2. Pour in 50 mL of freshly prepared staining solution (Subheading 2.6) and incubate for 45 min at room temperature in dark (*see* **Note 43**).
3. Remove the staining solution, wash the gel shortly twice with deionized water and transfer the gel on White tray (*see* **Note 39**).
4. Colored formazan produced by GSNOR dehydrogenase activity from NBT appear as purple-blue bands (*see* **Note 44**).
5. Place the tray inside a suitable documentation system, set the parameters for the illumination by a white light source and start imaging (*see* **Note 42**).

4 Notes

1. To avoid light-induced GSNO decomposition, all procedures with GSNO compound and solutions should be performed avoiding exposure to any light sources.
2. Reduced GSH stored at room temperature or at 4 °C undergoes a slow oxidation by the air oxygen. The use of old and inappropriately stored GSH results in decreased yield of the reaction product.

3. The reaction mixture will turn red, and intensive production of bubbles and eventually formation of red precipitate can be observed.
4. The precipitation yield and purity by subsequent wash steps are achieved using acetone and ether previously kept for 1 h at −20 °C.
5. The purity of synthesized GSNO can be verified by a measurement of absorbance at 334 nm in comparison to commercial GSNO sample (Calbiochem, San Diego, CA, USA) using the molar extinction coefficient for GSNO $\varepsilon = 800\ \mathrm{mM^{-1} \cdot cm^{-1}}$.
6. To inhibit proteases in the plant extract, AEBSF is recommended as a less toxic and more stable alternative to phenylmethylsulfonyl fluoride (PMSF). It can be purchased also under the commercial name Pefabloc (Roche Diagnostic).
7. The stock CBB solution should have reddish-brown color. It can be stored for several months in dark at 4 °C.
8. Freshly prepared GSNO solution should have a pink color. A GSNO stock solution can be prepared just prior to its use by dissolving GSNO in an organic solvent like DMSO purged with an inert gas. Aqueous GSNO solution should be prepared just prior to its use.
9. In contact with air, commercial concentrated formaldehyde solutions oxidize to formic acid and eventually polymerize to paraformaldehyde. It is therefore advisable to use formaldehyde from recently opened ampules. Alternatively, "fresh" formaldehyde solutions can be obtained by heating of paraformaldehyde.
10. The specificity of GSNOR activity staining can be checked by a parallel staining of another gel in a staining solution without glutathione addition.
11. Caution: Acrylamide is a neurotoxin. Always wear protective gloves, safety glasses, and a surgical mask when working with acrylamide powder. To avoid exposing your lab mates to acrylamide, work in the fume hood.
12. The use of freshly prepared ammonium persulfate solution is crucial for the initiation of acrylamide monomer polymerization to achieve the desired gel quality.
13. Running buffer for native PAGE can be reused several times, if stored at 4 °C. A stock solution of 10-times concentrated buffer can be alternatively stored at 4 °C and diluted prior use with cold deionized water.
14. Samples of smaller volume can be desalted on NAP-5 desalting columns (GE Healthcare, USA), loading 0.5 mL of extract supernatant and eluting with 1 ml of elution buffer.

15. Desalting columns are available as disposable consumables, but the experience in our lab shows that by taking a proper care, repeated uses are feasible without compromising the column efficiency. After elution step, wash the column with at least 25 mL of deionized water to clean the column before next desalting cycle. For long term storage, wash with at least 15 mL of 20% ethanol to avoid microbial contamination, and store at 4 °C.
16. Extract samples with higher amount of protein need to be diluted to fall with their concentration into the range of the calibration curve.
17. Prepared working Bradford reagent shall have bluish brown color.
18. A known drawback of the Bradford method is the long-term instability of color changes resulting from complexes formed with Coomassie Blue, namely in samples with high protein concentrations. For this reason, it is highly recommended to read absorbance values less than 20 min after the addition of the Bradford reagent to protein samples.
19. This modification of the Bradford assay reported by [16], based on using ratios of absorbance measurements at 595 and 450 nm rather than measurements at 595 nm only, provides an increased linearity of the function of absorbance ratios plotted against known concentrations of the calibration protein, usually bovine serum albumin.
20. EDTA or similar chelating agents should not be used neither during sample extraction nor during activity determination, as this can result in the removal of zinc atoms from GSNOR molecule and loss of the enzyme activity. Chelating agents like EDTA completely inactivate the enzyme in the millimolar range. Reducing agents like β-mercaptoethanol, dithiothreitol or ascorbic acid inhibit both the reductase and dehydrogenase reactions of GSNOR, and at 1 mM concentration they reduce the enzyme activity by about 30% [10].
21. A separate blank measurement must be performed for each sample. Ten to twenty microliters of desalted extract are usually optimal to measure GSNOR activity in plant extracts.
22. This time interval has been established as an optimum for measuring GSNOR activities in plant extracts. Analysis of samples with high GSNOR activity (i.e., purified recombinant proteins) will probably require shortened measuring time to obtain linear progress of the initial rate of the enzyme reaction.
23. For the calculation of concentration changes using Lambert–Beer equation, the optical path lengths (OPL) of the measured sample needs to be determined. If this option is not available on the used microplate reader, the height of the

solution in the microplate wells can be calculated using the known volume and the well surface, which should be available from the microplate manufacturer. Optical path length for each well sample can be determined by measurements of absorbance at 900 and 977 nm and calculations using formula OPL (cm) = $(A_{977}–A_{900})/0.18$.

24. Other GSNOR substrates are also used for the determination of dehydrogenase activity, namely long-chain alcohols as octanol or geraniol. In this case, the reaction buffer needs to have pH adjusted to 10.
25. *S*-(hydroxymethyl)glutathione (HMGSH) is an adduct of glutathione (GSH) and formaldehyde [17]. Glutathione and formaldehyde concentrations required for the preparation of 1 mM HMGSH were calculated using the equation and K_{eq} value as described previously [10, 18].
26. Preparation of two identical gels is suitable for the detection of GSNOR in one gel and parallel detection of total proteins by Coomassie Blue, silver staining or other available method in the other gel.
27. Buffers and acrylamide solution are usually stored at 4 °C. It is hence important to let the solution to equilibrate to the laboratory temperature before preparation of gel solutions, as a low temperature of the reaction solution is known to have negative effect on the acrylamide polymerization and the quality of resulting gels.
28. Acrylamide polymerization involves free-radical reactions, which are inhibited by oxygen or other compounds with free-radical scavenging properties. Solution degassing is therefore substantial to remove the molecular oxygen dissolved in the used solutions and buffers or adhered to the labware surfaces.
29. Acrylamide polymerization in the gel is initiated as soon as ammonium persulfate is added to the solution, so all subsequent actions must be performed promptly.
30. A butanol layer on the top of the gel limits the oxygen diffusion, thus accelerating gel polymerization.
31. Glycerol is mixed with desalted plants extracts to increase the density of sample solutions to enable their easy transfer to preformed wells in the stacking gel.
32. The electrode buffer provides an electric connection between the upper and lower ends of the gel assembly to generate an electrophoretic movement of proteins from the analyzed sample.
33. Air bubbles below the plates can result in irreproducible results due to a non-homogenous electric field and currents passing through the gels.

34. The used amount of bromophenol solution should be as low as possible; otherwise bromophenol can eventually diffuse to the lanes of adjacent samples and affect the activity of present proteins.
35. Electrophoresis power supplies should be used in accordance with general rules of using electric appliances. Never connect the lid and switch on the power supply before the lid is securely placed on the electrophoretic cell.
36. Cutting on of the gel corners will facilitate the recognition of a correct gel orientation in following steps of protein blotting.
37. For this purpose we successfully use papers available for the germination of plant seeds, as an economic alternative to commercial blotting papers provided by Bio-Rad or other suppliers.
38. In all manipulation steps with PAGE gels and blotting membranes, always wear quality protective gloves.
39. Alternatively, place gels on a piece of a clean glass, such as old glass plates for bigger gels.
40. The recommended volume is at least 2 mL for one piece of filter paper.
41. When using a glass plate as the gel support, place it on the top of a UV illuminator in the documentation system.
42. The exposure settings need to be optimised for each experiment and specific documentation system.
43. All solutions should be prepared fresh and protected from the light. Moreover, the staining solution should be thoroughly mixed after the addition of each compound. We recommend adding the compounds to the staining buffer strictly in the given order. If required, the method sensitivity can be increased by incubation of the gels with staining solution at 37 °C in dark.
44. The resulting color will depend on the choice of tetrazolium salt used for the staining. Generally, we have observed comparable results using NBT or MTT to detect GSNOR dehydrogenase activity. For some samples MTT provides more intense staining; this is however usually accompanied by a more intense background staining of the whole gel compared to NBT.

Acknowledgments

This project was supported by an internal Grant from Palacky University in Olomouc (IGA_PrF_2019_022).

References

1. Yu M, Lamattina L, Spoel SH, Loake GJ (2014) Nitric oxide function in plant biology: a redox cue in deconvolution. New Phytol 202:1142–1156
2. Seth D, Stamler JS (2011) The SNO-proteome: causation and classifications. Curr Opin Chem Biol 15:129–136
3. Martínez-Ruiz A, Lamas S (2004) S-nitrosylation: a potential new paradigm in signal transduction. Cardiovasc Res 62:43–52
4. Corpas FJ, Alché JD, Barroso JB (2013) Current overview of S-nitrosoglutathione (GSNO) in higher plants. Frontiers Plant Sci 4:126
5. Benhar M, Forrester MT, Stamler JS (2009) Protein denitrosylation: enzymatic mechanisms and cellular functions. Nat Rev Mol Cell Biol 10:721–732
6. Koivusalo M, Baumann M, Uotila L (1989) Evidence for the identity of glutathione-dependent formaldehyde dehydrogenase and class III alcohol dehydrogenase. FEBS Lett 257:105–109
7. Jensen D, Belka G, Du Bois G (1998) S-Nitrosoglutathione is a substrate for rat alcohol dehydrogenase class III isoenzyme. Biochem J 331:659–668
8. Liu L, Hausladen A, Zeng M et al (2001) A metabolic enzyme for S-nitrosothiol conserved from bacteria to humans. Nature 410:490–494
9. Staab CA, Hellgren M, Höög JO (2008) Medium- and short-chain dehydrogenase/reductase gene and protein families: dual functions of alcohol dehydrogenase 3: implications with focus on formaldehyde dehydrogenase and S-nitrosoglutathione reductase activities. Cell Mol Life Sci 65:3950–3960
10. Kubienová L, Kopečný D, Tylichová M et al (2013) Structural and functional characterization of a plant S-nitrosoglutathione reductase from *Solanum lycopersicum*. Biochimie 95:889–902
11. Leterrier M, Chaki M, Airaki M et al (2011) Function of S-nitrosoglutathione reductase (GSNOR) in plant development and under biotic/abiotic stress. Plant Signal Behav 6:789–793
12. Xu S, Guerra D, Lee U, Vierling E (2013) S-nitrosoglutathione reductases are low-copy number, cysteine-rich proteins in plants that control multiple developmental and defense responses in *Arabidopsis*. Frontiers Plant Sci 4:430
13. Corpas FJ, Carreras A, Esteban FJ et al (2008) Localization of S-nitrosothiols and assay of nitric oxide synthase and S-nitrosoglutathione reductase activity in plants. Methods Enzymol 437:561–574
14. Chaki M, Valderrama R, Fernández-Ocaña AM et al (2011) Mechanical wounding induces a nitrosative stress by down-regulation of GSNO reductase and an increase in S-nitrosothiols in sunflower (*Helianthus annuus*) seedlings. J Exp Bot 62:1803–1813
15. Barnett SD, Buxton LO (2017) The role of S-nitrosoglutathione reductase (GSNOR) in human disease and therapy. Crit Rev Biochem Mol Biol 52:340–354
16. Zor T, Selinger Z (1996) Linearization of the Bradford protein assay increases its sensitivity: theoretical and experimental studies. Anal Biochem 236:302–308
17. Moore KP, Mani AR (2002) Measurement of protein nitration and S-nitrosothiol formation in biology and medicine. In: Cadenas E, Lester P (eds) Methods in enzymology. Academic Press, Cambridge, MA, pp 256–268
18. Uotila L, Koivusalo M (1974) Purification and properties of S-formylglutathione hydrolase from human liver. J Biol Chem 249:7664–7672

Chapter 6

Expression Analysis of Important Genes Involved in Nitrogen Metabolism Under Hypoxia

Mallesham Bulle, Reddy Kishorekumar, Aakanksha Wany, and Kapuganti Jagadis Gupta

Abstract

Hypoxia or anoxia condition can occurs during flooding or waterlogging and can cause intense damage to the plants. Since oxygen is important for active operation of electron transport chain in mitochondria to generate energy production (ATP) any drop in oxygen can cause an energy crisis during flooding/waterlogging. To cope with this energy crisis plants have developed various anatomical, physiological, and biochemical adaptations. Perception of signals and induction of genes are required for initiation of these adaptive responses. Various genes involved in nitrogen, carbon, and fermentative metabolism play a role in hypoxic tolerance. Regulation of genes involved in nitrogen metabolism also plays a role under hypoxia. Hence in this present chapter we describe the expression of nitrate reductase-1 (*NIA1*), nitrate reductase-2 (*NIA2*), and glutamine synthetase-1 (*GLN-1*) during hypoxia in Arabidopsis.

Key words Nitrogen, Nitrate reductase 1 (*NIA1*), Nitrate reductase 2 (*NIA2*), Glutamine synthetase 1 (*GLN1*)

1 Introduction

Nitrogen (N) is requirement for plant growth, development, and resistance against various biotic abiotic stresses. Numerous processes are mediated by N including synthesis of amino acids, secondary metabolites, nucleic acids, vitamins, and operation of various primary and defense regulatory pathways. N composition is not the same in all soils. It varies greatly, and environmental factors play an important a role in determining N concentration and its form in the soil [1]; thus, nitrogen concentration indeed plays a role in resistance against stress [2].

Plants are equipped with a wide array of sophisticated mechanisms for sensing, uptake, and translocation of N in to their cells for assimilation as per cellular requirements. Nitrate is taken up by the roots. Once it is taken up it is reduced to nitrite or ammonium or stored in vacuoles for the purpose of osmoregulation [3]. Nitrate is

Kapuganti Jagadis Gupta (ed.), *Nitrogen Metabolism in Plants: Methods and Protocols*, Methods in Molecular Biology, vol. 2057, https://doi.org/10.1007/978-1-4939-9790-9_6,

also rapidly transported to the leaves where it is reduced to ammonium, whereas ammonium directly taken up by the roots is assimilated via the GS/GOGAT pathway. The operation of this pathway leads to formation of glutamine/glutamate which then acts as a substrate for amino acid biosynthesis via transamination reactions [4].

Plant metabolism is affected by nitrogen concentration, such that optimum nitrogen concentration leads to active operation of the primary metabolism, leading to higher levels of proteins and faster growth [5, 6]. Low levels of N can cause repressive effects on growth and can compromise plant defense response to stress [5, 6].

As mentioned above nitrogen metabolism plays an important role under stress conditions as well. For instance under low oxygen conditions genes and enzymes of nitrogen metabolism are suppressed under ammonium form of nutrition in comparison to nitrate form of nutrition [7]. In response to low oxygen availability, plant induces differential gene expression involved in primary carbon and nitrogen metabolism [8] and this strongly effects induction of fermentations pathways such as alcohol and lactate fermentative pathways [7, 9] to generate ATP under hypoxia [10, 11]. This process was demonstrated in various species such as rice, Arabidopsis, and Populus [12]. Due to requirement of oxygen any depletion of oxygen via hypoxia/flooding inhibits the mitochondrial oxidative phosphorylation leads to reduced respiration via cytochrome pathway and induction of alternative pathway [13]. Under such circumstances nitrate reduction and nitric oxide production helps in generation of energy via phytoglobin nitric oxide cycle thus considered as alternative routes of fermentation for production of limited amount of energy [14]. Previously it was shown that both hypoxia and anoxia induce the expression of several key genes involved in N metabolism pathway such as nitrate reductase genes *NIA1,* Nitrate/chlorate transporters, and glutamate decarboxylase in Arabidopsis [15]; glutamine synthetase in rice [16]; and alanine aminotransferase in barley [17] and soy bean [18]. Root tissues exposed to hypoxia-induced cytosolic NR greatly suppressed the nitrite reductase activity [7, 19–21]. In *Arabidopsis* root cultures, NR genes are hypoxically inducible. *NR1* gene was upregulated after 2–4 and 20 h of hypoxic treatments [15]. A fivefold induction of *NIA*1 transcript levels were reported in Arabidopsis under nitrate nutrition [7]. The expression of *GDH1* isoforms enhanced during the different hypoxic conditions but only slight elevation of these transcripts were observed in *GDH3* in *Medicago*, however the enzyme activity remained lower than the normoxic conditions [22]. These authors speculated that the GDH expression during hypoxia could be a priming response which can assist during post hypoxia recovery [22]. Accumulation of several metabolites such as Alanine and ɣ aminobutyric acid (GABA) were also observed [8, 18]. During hypoxia, NADH-GOGAT activity in roots can regenerate NAD^+, to help in maintenance of glycolysis [22]. Under hypoxic conditions the accumulated pyruvate is converted to 2-oxoglutarate via AlaAT reaction, mobilizes to TCA cycle and used by oxoglutarate dehydrogenase (OGDH) and succinate CoA ligase to generate some ATP. The required NAD^+ for reduction of 2-OG can be obtained by increased activity of malate dehydrogenase (MDH) [8]. Hypoxic conditions activates the fermentation pathway related genes such as lactate dehydrogenase, alcohol dehydrogenase, and pyruvate decarboxylase [23]. The operation of fermentative pathways help in generation of more ATP via split TCA linked to nitrogen metabolism via AlaAT. Since plants adaptive response requires ATP, the coordination of glycolysis, TCA cycle, and nitrogen metabolism helps in hypoxic tolerance in plants.

2 Materials

2.1 Plant Material

Arabidopsis thaliana (Columbia-0) seedlings (14 days old).

2.2 RNA Isolation, cDNA Synthesis, and Quantitative Real-Time PCR

2.2.1 Reagents/Chemicals

1. Tri Reagent (MRC, USA) or TRIzol (RNA IsoPlus, TaKaRa).
2. Liquid nitrogen.
3. Chloroform (*see* **Note 1**).
4. Prechilled isopropanol.
5. Prechilled 3 M Sodium acetate (pH 5.2).
6. 70% DEPC ethanol
7. DEPC water (nuclease free-water) (*see* **Note 1**).
8. Triton X-100.
9. First strand cDNA synthesis kit (Fermentas, USA).
10. SYBR Green Master Mix (Applied Biosystems, USA).
11. Diluted cDNA template (25–100 ng/μl).
12. Forward Primer (*see* Table 4).
13. Reverse Primer (*see* Table 4).
14. Murashige and Skoog media (PT021, Himedia, India).

2.2.2 Miscellaneous Items

1. 1.5 ml microcentrifuge tubes (Eppendorf).
2. Sterile mortar and pestle.
3. 0.2 ml PCR tubes.
4. DEPC-treated tips (*see* **Note 2**).

2.3 Equipment

1. Hypoxia chamber (Bioxia, IMSET, India) with Nitrogen and oxygen gas cylinders.
2. Centrifuge.

3. NanoDrop 1000 spectrophotometer (Thermo Scientific, USA).
4. Thermal cycler (Applied Biosystems, USA).
5. 96-well qPCR plates (Applied Biosystems, USA)
6. ABI Prism 7900 sequence detection system (Applied Biosystems, USA).
7. Growth chamber.

3 Methods

3.1 Plant Growth and Conditions

1. Surface sterilize the Arabidopsis seeds (~100 seeds) by adding 400–500 μl of Triton X-100 + 75% ethanol after that keep the vials in a shaker for 5 min at room temperature.
2. After shaking replace the above solution by equal amount of 100% ethanol and keep in shaker under the same condition for another 2–3 min.
3. Remove the seeds over sterile filter paper under the laminar air flow using a sterile micro tip, and keep them for some time until all the seeds are dried.
4. Transfer 15–20 sterile seeds in a row with sterile toothpick to the plates containing 0.5× MS media.
5. Keep the inoculated seed plates for stratification in the cold room (4 °C) for 2 days in dark.
6. After 2 days, transfer the plates to the growth chamber for 14 days at a photoperiod of 8/16 h light/dark cycle, day/night temperature regime of 23°/18 °C, RH- 60% and photon flux density of 180–200 μmol m^{-2} s^{-1}.

3.2 Treatments

3.2.1 Hypoxia Treatment

1. Incubate 14-day-old plants in Bioxia chamber set at 0.4% oxygen for 3 h (Fig. 1).
2. Immediately freeze samples after hypoxia in liquid nitrogen.

3.2.2 Light Treatment

1. Grow 14-day-old plants under continuous light using Phillips TLD 36 W 1/830 light bulbs (150 μmol m^{-2} s^{-1} 188 PAR) at 23 °C for 3 h.
2. Shock freeze samples after hypoxia in liquid nitrogen.

3.3 RNA Isolation

1. Grind 500 mg freshly harvested plant material to a fine powder with liquid nitrogen (*see* **Note 3**).
2. To this powder, add 600–800 μl of TRIzol reagent and homogenize for 15 min at room temperature using a sterile mortar and pestle.

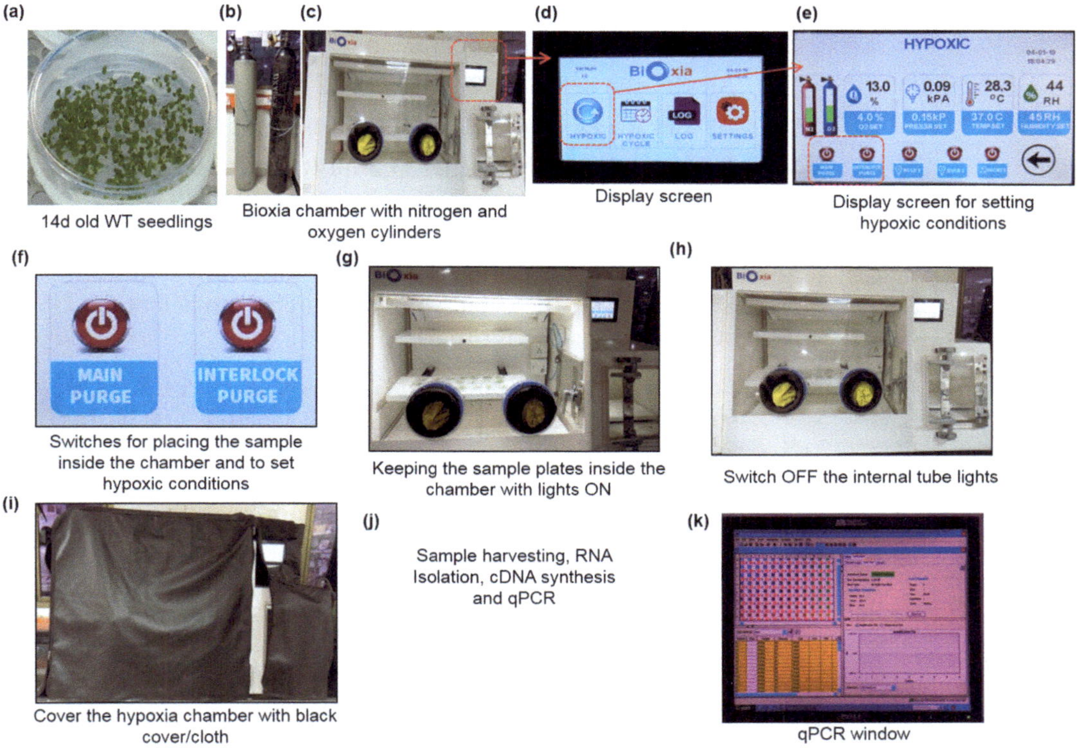

Fig. 1 Hypoxia treatment using BioXia-800 chamber and analysis of gene expression: Various steps are described in the figure starting from growth of plant material to analysis of gene expression

3. Transfer the homogenate to a fresh Eppendorf tube to this add 300 μl chloroform and vortex gently (and frequently) for 12–15 min at room temperature.
4. Centrifuge this mixture for a period of 15 min at 14,500 × *g* at 4 °C.
5. Transfer supernatant to fresh 1.5 ml microcentrifuge tubes and again, add 400 μl of chloroform (*see* **Note 4**).
6. Centrifuge for 10 min at 14,500 × *g* at 4 °C.
7. Gently recover the aqueous phase to a fresh 1.5 ml microcentrifuge tube.
8. Add 300 μl prechilled isopropanol and 100–150 μl pre-chilled 3 M sodium acetate.
9. Mix gently and incubate at −20 °C for 3 to 6 h or overnight.
10. After incubation centrifuge for 20 min at 16,800 × *g* at 4 °C and discard supernatant.
11. Wash the pellet with 75% ethanol for 5 min at 16,800 × *g* at 4 °C in a centrifuge.

12. Repeat the pellet washing step again with 500 μl of 75% ethanol, and discard the supernatant completely.
13. Air-dry the pellet for 2–3 min at 37 °C set in a thermoblock.
14. Air-dry it again for 5–10 min with open lid.
15. Resuspend the RNA pellet in 20μl RNAse-free water.
16. Quantify RNA in NanoDrop spectrophotometer.

3.4 cDNA Synthesis

Follow the protocol given in the kit used for cDNA synthesis. Here we used cDNA synthesis kit from Verso (Thermo Scientific, USA). Follow these steps (*see* **Note 5**).

1. For a single reaction (20 μl), in a 0.2 ml PCR tube, add stepwise the ingredients in the Table 1 below: (*see* **Note 6**).
2. Program the conditions in a PCR machine as described below in Table 2:
3. Store the synthesized cDNA in 4 °C or −20 °C for prolonged storage.

3.5 Quantitative Real-*Time PCR*

1. Set up a 96/384-well optical plate.
2. Prepare a PCR mixture (20 μl) using the following proportions per reaction in Table 3:

Table 1
Components, reagents required for synthesis of cDNA

Ingredients	Volume	Final concentration
5× cDNA synthesis buffer	4 μl	1×
dNTP mix	2 μl	500 μM each
RNA primer	1 μl	
RT enhancer	1 μl	
Verso enzyme mix	1 μl	
RNA template	1–5 μl	1 μg
Water, nuclease-free	To 20 μl	
Total volume	20 μl	

Table 2
Program and the conditions for setting in PCR for cDNA synthesis

	Temperature	Time	Number of cycles
cDNA synthesis	42 °C	30 min	1 cycle
Inactivation	95 °C	2 min	1 cycle

Table 3
Components used in each well for real-time PCR

Ingredients	Volume
SYBR green dye	10 μl
Forward primer	0.5 μl
Reverse primer	0.5 μl
cDNA template (100 ng)	1 μl
Water, nuclease-free	To 20 μl
Total volume	20 μl

3. Tightly seal the 96-well plate with the qPCR sealing film.
4. Centrifuge the plate briefly to eliminate any air bubbles in wells.
5. Place the samples in thermal cycler ABI Prism 7900 sequence detection system (Applied Biosystems, USA) programmed for the amplification:

   95 °C for 2 min, 40 cycles at 95 °C for 30 s, 60 °C for 30 s, 72 °C for 30 s; 60 °C for 15 s and 95 °C for 15 s.
6. Calculate fold expression values according to $2^{\Delta\Delta Ct}$ method (Fig. 2).
7. Use internal reference gene (18sRNA) for data normalization (Table 4).
8. Prepare graphs of relative expression values w.r.t. normoxic conditions.

4 Notes

1. DEPC, chloroform, ethanol, and ethidium bromide are very hazardous and must be handled with caution.
2. Use DEPC-treated pipette tips, PCR tubes, mortar and pestle, and microcentrifuge tubes.
3. Always use ethanol-cleaned area of the laboratory for RNA work.
4. Keep the aliquots on ice.
5. Do not vortex the Verso Enzyme Mix.
6. Use good-quality RNA, and it should be free of DNA and nucleases.

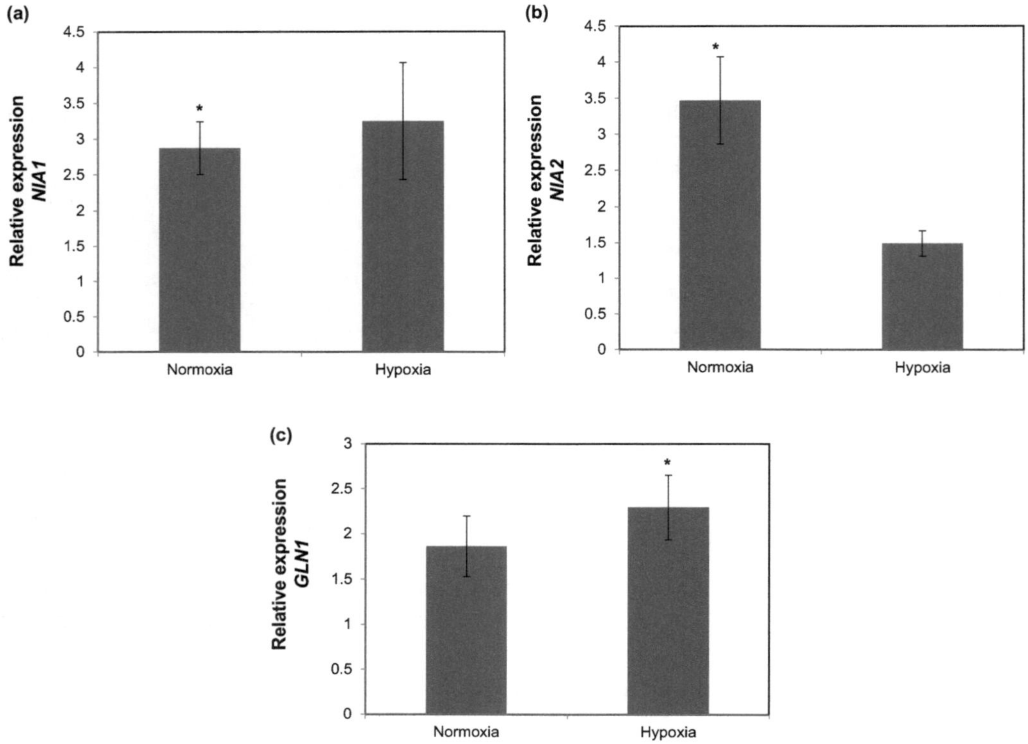

Fig. 2 Relative expression of genes involved in nitrogen metabolism: (**a**) *NIA1 (NITRATE REDUCTASE1; AT1G77760)*, (**b**) *NIA2 (NITRATE REDUCTASE2; AT1G37130)*, (**c**) *Gln1 (GLUTAMINE SYNTHETASE-1; AT1G48470)*. Values are means ($n = 3$) $\pm$ SE. Asterisk ($*$) indicates that statistical significance between light and dark treatments is evaluated by *t*-test at $p < 0.05$ ($*$)

Table 4
List of primers used in real time PCR

Gene	Accession number	Primer sequence (5′-3′)
NITRATE REDUCTASE1 (*NIA1*)	AT1G77760	F-TATGGGAAGTTCTGGTGTTG R-CTTCACTCTAAACCAGCAGT
NITRATE REDUCTASE2 (*NIA2*)	AT1G37130	F-TATGGGAAGTTCTGGTGTTG R-CTTCACTCTAAACCAGCAGT
GLUTAMINE SYNTHETASE-1 (*GLN1*)	AT1G48470	F-TTACTGAGATATGCGGTGTG R-CTTCACCATAAGCAGCAATG
18SrRNA	GQ380689	F-TCCTAGTAAGCGCGAGTCATCA R-CGAACACTTCACCGGATCAT

References

1. Stitt M (1999) Nitrate regulation of metabolism and growth. Curr Opin Plant Biol 2:178–186
2. Gupta KJ, Brotman Y, Segu S, Zeier T, Zeier J, Persijn ST, Cristescu SM, Harren FJ, Bauwe H, Fernie AR, Kaiser WM (2012) The form of nitrogen nutrition affects resistance against *Pseudomonas syringae pv. phaseolicola* in tobacco. J Exp Bot 64:553–568
3. Tischner R (2000) Nitrate uptake and reduction in higher and lower plants. Plant Cell Environ 23:1005–1024
4. Crawford NM, Forde BG (2002) Molecular and developmental biology of inorganic nitrogen nutrition. Arabidopsis Book 1. https://doi.org/10.1199/tab.0011
5. Roycewicz P, Malamy JE (2012) Dissecting the effects of nitrate, sucrose and osmotic potential on Arabidopsis root and shoot system growth in laboratory assays. Philos Trans R Soc B 367:1489–1500
6. Landrein B, Formosa-Jordan P, Malivert A, Schuster C, Melnyk CW, Yang W, Turnbull C, Meyerowitz EM, Locke JC, Jönsson H (2018) Nitrate modulates stem cell dynamics in Arabidopsis shoot meristems through cytokinins. Proc Natl Acad Sci U S A 115:1382–1387
7. Wany A, Gupta AK, Kumari A, Mishra S, Singh N, Pandey S, Vanvari R, Igamberdiev AU, Fernie AR, Gupta KJ (2018) Nitrate nutrition influences multiple factors in order to increase energy efficiency under hypoxia in Arabidopsis. Ann Bot. https://doi.org/10.1093/aob/mcy202
8. Rocha M, Licausi L, Araújo WL, Nunes-Nesi A, Sodek L, Fernie AR, Dongen JT (2010a) Glycolysis and the TCA-cycle are linked by alanine aminotransferase during hypoxia induced by waterlogging of *Lotus japonicus*. Plant Physiol 152:1501–1513
9. Bailey-Serres J, Voesenek LACJ (2008) Flooding stress: acclimations and genetic diversity. Annu Rev Plant Biol 59:313–339
10. Greenway H, Gibbs J (2003) Mechanisms of anoxia tolerance in plants. II. Energy requirements for maintenance and energy distribution to essential processes. Funct Plant Biol 30:999–1036
11. Geigenberger P (2003) Response of plant metabolism to too little oxygen. Curr Opin Plant Biol 6:247–256
12. Narsai R, Rocha M, Geigenberger P, Whelan J, van Dongen JT (2011) Comparative analysis between plant species of transcriptional and metabolic responses to hypoxia. New Phytol 190:472–487
13. Gupta KJ, Zabalza A, Van Dongen JT (2009) Regulation of respiration when the oxygen availability changes. Physiol Plant 137:383–391
14. Stoimenova M, Igamberdiev AU, Gupta KJ, Hill RD (2007) Nitrite-driven anaerobic ATP synthesis in barley and rice root mitochondria. Planta 226:465–474
15. Klok EJ, Wilson IW, Wilson D, Chapman SC, Ewing RM, Somerville SC, Peacock WJ, Dolferus R, Dennis ES (2002) Expression profile analysis of the low-oxygen response in Arabidopsis root cultures. Plant Cell 14:2481–2494
16. Reggiani R, Nebuloni M, Mattana M, Brambilla I (2000) Anaerobic accumulation of amino acids in rice roots: role of the glutamine synthetase/glutamate synthase cycle. Amino Acids 18:207–217
17. Good AG, Crosby WL (1989) Anaerobic induction of alanine aminotransferase in barley root tissue. Plant Physiol 90:1305–1309
18. Rocha M, Sodek L, Licausi F, Hameed MW, Dornelas MC, van Dongen JT (2010b) Analysis of alanine aminotransferase in various organs of soybean (*Glycine max*) and in dependence of different nitrogen fertilisers during hypoxic stress. Amino Acids 39:1043–1053
19. Botrel A, Kaiser WM (1997) Nitrate reductase activation state in barley roots in relation to the energy and carbohydrate status. Planta 201:496–501
20. Botrel A, Magne C, Kaiser WM (1996) Nitrate reduction, nitrite reduction and ammonium assimilation in barley roots in response to anoxia. Plant Physiol Biochem 34:645–652
21. Planchet E, Jagadis Gupta K, Sonoda M, Kaiser WM (2005) Nitric oxide emission from tobacco leaves and cell suspensions: rate limiting factors and evidence for the involvement of mitochondrial electron transport. Plant J 41:732–743
22. Limami AM, Glevarec G, Ricoult C, Cliquet JB, Planchet E (2008) Concerted modulation of alanine and glutamate metabolism in young *Medicago truncatula* seedlings under hypoxic stress. J Exp Bot 59:2325–2335
23. Kennedy RA, Rumpho ME, Fox TC (1992) Anaerobic metabolism in plants. Plant Physiol 100:1–6

Chapter 7

Enzymatic Conversions of Glutamate and γ-Aminobutyric Acid as Indicators of Plant Stress Response

Alexander T. Eprintsev, Natalia V. Selivanova, and Abir U. Igamberdiev

Abstract

Glutamate plays a central role in amino acid metabolism, in particular, in aminotransferase reactions leading to formation of many other proteinogenic and nonproteinogenic amino acids. In stress conditions, glutamate can be either metabolized to γ-aminobutyric acid (GABA) by glutamate decarboxylase which initiates a GABA shunt bypassing several reactions of the tricarboxylic acid cycle or converted to 2-oxoglutarate by glutamate dehydrogenase. Both reactions direct protein carbon to respiration but also link glutamate metabolism to other cellular pathways, resulting in the regulation of redox level and pH balance. Assays for determination of activities of glutamate dehydrogenase and of the GABA shunt enzymes as the markers of stress response is described in this chapter. These assays are important in the studies of the strategy of biochemical adaptation of plants to changing environmental conditions including elevated CO_2, temperature increase, flooding, and other stresses.

Key words Amino acid metabolism, γ-aminobutyric acid (GABA), Glutamate, Respiration, Stress

1 Introduction

The mechanism of plant response to stress includes modulation of cellular biochemical processes constituting the oxidative branch of the primary and secondary metabolism [1]. Metabolic conversions of free amino acids occupy a major place in stress response, while glutamate plays a key role in these conversions [2–4]. Physiological state of plants can be estimated based on the levels and changes of activities of the enzymes controlling these conversions. Stress metabolites and the enzymes producing them can serve as biochemical indicators of physiological state of plants in changing environmental conditions. Among the stress metabolites derived from glutamic acid, in addition to such amino acids as proline and arginine [5, 6], γ-aminobutyric acid (GABA) represents the most abundant nonproteinogenic amino acid which serves as a reserve of reduced nitrogen that is not used directly for protein synthesis and

Kapuganti Jagadis Gupta (ed.), *Nitrogen Metabolism in Plants: Methods and Protocols*, Methods in Molecular Biology, vol. 2057, https://doi.org/10.1007/978-1-4939-9790-9_7,

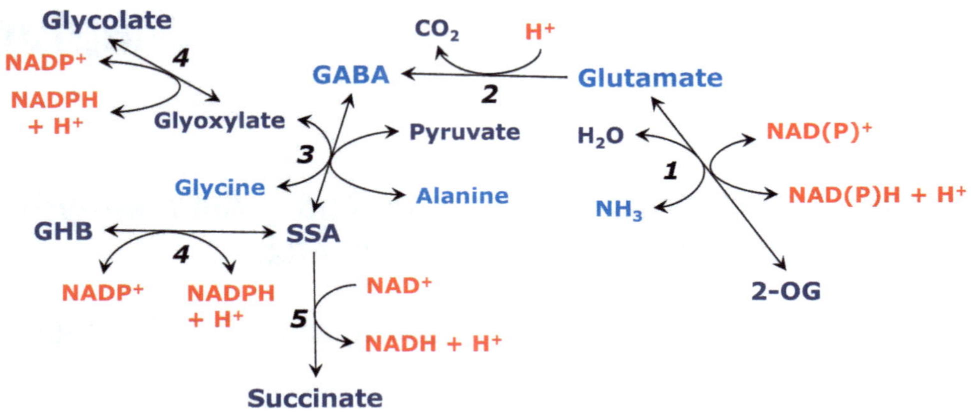

Fig. 1 Enzymatic conversions of glutamate and γ-aminobutyric acid (GABA) in plants. Abbreviations: *2-OG* 2-oxogutarate, *SSA* succinic semialdehyde, *GHB* γ-hydroxybutyrate. Enzymes: 1—glutamate dehydrogenase; 2—glutamate decarboxylase; 3—GABA transaminase; 4—succinic semialdehyde/glyoxylate reductase; 5—succinic semialdehyde dehydrogenase

represents the less abundant nitrogen pool as compared to the pool of proteinogenic amino acids [7–9].

GABA was first detected in plants in 1940s [10] and then attracted much interest because of the increase of its levels under different kinds of abiotic stresses [2, 4]. The increase of GABA content in plant tissues is observed under the influence of anaerobiosis, flooding, drought, impaired mineral nutrition, low temperature, radiation, wounding, low light, and low pH [8, 11, 12]. It has been shown that GABA serves as a modulator of ionic transport in plants and thus participates in the transmission of stress signals [7]. GABA is produced from glutamate by glutamic acid decarboxylase (GAD; EC 4.1.1.15), the reaction consuming proton and releasing CO_2, while glutamate dehydrogenase (GDH; EC 1.4.1.2) catalyzes the alternative reaction of conversion of glutamate to 2-oxoglutarate using NAD^+ or $NADP^+$ as a cofactor (Fig. 1).

GABA is transaminated by the enzyme GABA transaminase (GABA-T; EC 2.6.1.19) into succinic semialdehyde (SSA) which can be further converted to succinate by SSA dehydrogenase (SSADH; EC 1.2.1.24). Succinate then enters the tricarboxylic acid (TCA) cycle [13]. Alternatively, SSA is reduced to γ-hydroxybutyrate (GHB) by γ-hydroxybutyrate dehydrogenase (GHBDH; EC 1.1.1.61) which is accumulated under stress [14]. Under oxygen deficiency, SSADH activity is limited by the increased reduction level, inhibition by adenylates and low pH, which facilitates GABA accumulation [15, 16]. In these conditions, SSA is metabolized to GHB by GHBDH which acts as SSA reductase [17, 18]. GHBDH was detected in plant tissues where it is induced in response to oxygen deficiency [14, 19]. Metabolism of

SSA to GHB is considered a type of fermentation in addition to the fermentative reactions catalyzed by lactate dehydrogenase and alcohol dehydrogenase [20]. Thus GHBDH can participate in the protection of plants against oxidative stress [21].

Due to the coordinated action of all above mentioned enzymes, the incorporation of amino groups into different amino acids takes place and simultaneously such parameters as pH, energy state, and redox level are regulated. Since GABA shunt and related processes play an important role in bioenergetics of stressed plants, the measurement of GDH, GAD, GABA-T, SSADH, and GHBDH becomes important for tracing metabolic changes occurring in plants under the influence of different types of abiotic stress.

2 Materials

The chemicals for measurement of enzyme activities could be obtained from Sigma-Aldrich or other suppliers. The instrumentation includes any refrigerated centrifuge and UV-Visible spectrophotometer. In some cases HPLC and scintillation counter can be used, however, the methods that involve this instrumentation are supplementary to the simple spectrophotometric methods of measurement of enzyme activities. Preparation of enzyme extracts can be performed by grinding of plant samples in liquid nitrogen using mortar and pestle or a homogenizer and then extracting the powder to ice-cold 100 mM Tris–HCl buffer, pH 8.0, containing 1 mM EDTA, 2 mM dithiothreitol, and 2 mM $MgCl_2$. It is recommended to add 1% (v/v) cocktail of protease inhibitors and 10% (v/v) glycerol for stabilization of the studied enzymes. The composition of buffer and its pH value can vary depending on the studied enzyme, however, in all cases the range of pH 7.5–8 is suitable for extraction. After centrifugation at 10,000 × *g* for 10 min at 4 °C the supernatant is used for analysis of enzyme activities and for quantitative estimation of total protein. For more precise measurements the supernatant can be subjected for gel filtration on the column packed with Sephadex G-10, G-25, or a similar sorbent material. In advanced studies the fractionation of cellular compartments (mitochondria, cytosol) using ultracentrifuge and gradients of Percoll or sucrose precedes enzyme measurements. This is not described in the current chapter and is available in the literature.

3 Methods

3.1 Measurement of Glutamate Dehydrogenase Activity

Many different aminotransferases (transaminases) use 2-oxoglutarate as amino group acceptor forming glutamate as a product. Glutamate further can be either oxidized with the release of ammonia or decarboxylated (*see* **Note 1**). Oxidative deamination of glutamate is catalyzed by GDH which usually represents a hexamer having a molecular weight of more than 300 kDa, localized in mitochondria and can use NAD^+ or $NADP^+$ as a coenzyme [22]. The action of this enzyme completes preceding deamination of many amino acids, while ammonia released in this reaction can be used for urea synthesis and other processes [23].

The reaction of GDH can be measured in the direction of glutamate conversion in 100 mM Tris–HCl or bicarbonate buffer, pH 8.1, containing 1 mM glutamate and 1.5 mM NAD^+. The reverse reaction needs high concentrations of ammonium and therefore is considered unfavorable in physiological conditions, although in the conditions of high ammonium supply it can provide a marked rate of glutamate synthesis [24]. To measure the reaction of glutamate formation, it is recommended to use 100 mM Tris–HCl buffer, pH 7.5, containing 2.5 mM 2-oxoglutarate, 0.2 mM NADH and 50 mM ammonium chloride. Both reactions are monitored spectrophotometrically at 340 nm [25, 26] and the rate is calculated using the extinction coefficient of NADH and NADH of 6.22 $mmol^{-1}$ cm^{-1}.

3.2 Measurement of Glutamate Decarboxylase Activity

GAD catalyzes the conversion of glutamic acid to GABA and CO_2 according to the following equation:

$$^{-}OOC - CH_2 - CH_2 - CH(NH_3{}^+) - COO^- + H^+ \\ \rightarrow CO_2 + {}^{-}OOC - CH_2 - CH_2 - CH_2 - NH_3{}^+$$

It has been characterized in animals, plants, yeast and bacteria [27, 28].

Activity of GAD in plants can be measured at pH 6.8 which represents average cytosolic pH value under hypoxic conditions, although pH optimum of GAD can be lower. For the assay medium, 50 mM potassium phosphate buffer, pH 6.8, containing 0.1 mM pyridoxal phosphate, 0.1 mM dithiothreitol and 5 mM L-glutamate can be used. After the incubation for 30 min, the reaction is stopped by boiling for 10 min to inactivate the enzyme. Prior to boiling some authors recommend adding the equal volume of 0.2 M borate, pH 10.0 to achieve fast stopping the reaction [29]. The formed GABA can be quantified either by HPLC or by the method described in Park et al. [30] using the preparation of GABase (Sigma) which represents a mixture of GABA transaminase and SSADH (*see* **Note 2**). One milliliter assay mixture consists of 0.4 M potassium pyrophosphate buffer, pH 8.6, 0.6 mM $NADP^+$, 0.05 units of GABase (Sigma), and 1 mM 2-oxoglutarate. The initial absorbance is read at 340 nm before adding 2-oxoglutarate, and the final absorbance is read after 60 min incubation at 25 °C. The difference used to calculate GABA content in the sample using the equation of the GABA standard curve: $A_{340} = 0.0062X - 0.0027$, where X = concentration of GABA in mM. One unit of GAD activity is defined as the amount of enzyme producing 1 μmol GABA per minute.

The method of GAD assay using radioactive glutamate and trapping $^{14}CO_2$ [31] has some advantages because it does not require the additional step of measuring GABA, however, its application is limited by availability of the isotope lab. This method can be modified for plants by using 50 mM potassium phosphate buffer, pH 6.8, containing 0.1 mM pyridoxal phosphate, 0.1 mM dithiothreitol and 5 mM L-glutamate containing 0.02 μCi ^{14}C-glutamate. The released $^{14}CO_2$ is trapped by filter paper fixed by needle on the cork top of the assay tube and soaked with either 1 M NaOH or with monoethanolamine. After the period of incubation (1 h or 30 min) the filter paper is removed and placed in the scintillation fluid for counting. The GAD activity is calculated taking into account the incubation time, protein concentration and specific radioactivity per μmol glutamate. Activation of GAD by calcium and calmodulin can be a matter of a separate study while adding of Ca^{2+} is not necessary for GAD assay that reflects the existing activation level of this enzyme in the cell.

3.3 Measurement of Succinic Semialdehyde Dehydrogenase Activity

SSADH is the enzyme that is most active at alkaline pH values. The bifurcation of SSA transition either to GHB or to succinate can be controlled by pH values in the mitochondrial matrix. For the extraction of SSADH 25 mM sodium phosphate buffer (pH 9.0) containing 1 mM DTT can be used. The standard assay of this enzyme can be performed in 0.1 M sodium phosphate buffer, pH 9.0, containing 1 mM DTT, 0.1 mM SSA and 0.5 mM NAD^+. Spectrophotometric detection of the formed NADH is monitored at 340 nm [14].

3.4 Measurement of GABA Transaminase Activity

Spectrophotometric determination of GABA-T activity can be performed in 50 mM Tris–HCl buffer, pH 8.0, containing 1.5 mM DTT, 0.75 mM EDTA, 0.1 mM pyridoxal-5-phosphate (PLP), 10% (v/v) glycerol, 16 mM GABA and 4 mM pyruvate. The volume of 0.15 ml should contain ~30 μl protein. Control sample contains boiled enzyme instead of the native enzyme extract. After the incubation at 30 °C for 60 min the samples are placed to water bath at 97 °C for 7 min to stop the reaction. The GABA-T activity is then estimated enzymatically via quantitative determination of the product L-alanine by using alanine dehydrogenase (AlaDH). For this purpose, 40 μl sample is added to 160 μl of sodium carbonate buffer, pH 10.0, containing 1 mM NAD^+ and 0.02 units of AlaDH from *Bacillus subtilis* (Sigma) (*see* **Note 2**) and the increase in optical density at 340 nm is measured. The value of optical density obtained by using the control sample with boiled enzyme should be subtracted. The amount of L-alanine is calculated from the calibrating curve prepared with different concentrations of L-alanine [32].

3.5 Measurement of γ-Hydroxybutyrate Dehydrogenase Activity

GHBDH from *Arabidopsis thaliana* does not show significant homology with GHBDH from bacteria and animals, but its sequence shows similarity to several plant dehydrogenases including β-hydroxyacid dehydrogenases and 6-phosphogluconate dehydrogenase [14]. Spectrophotometric analysis of GHBDH activity in the direction of SSA reduction can be performed in 100 mM MOPS-KOH buffer, pH 7.2, containing 5 mM SSA, 0.25 mM NADPH, 10 mM $MgCl_2$, 10 mM KCl, and 1 mM $MnCl_2$. For measurement of the reaction in the direction of GHB oxidation, SSA and NADPH should be replaced by 5 mM GHB and 1 mM $NADP^+$, and 100 mM Tris–HCl buffer, pH 8.8 should be used. The reactions are initiated by adding SSA or GHB correspondingly and controlled by measuring of change in NADPH concentration at 340 nm [33].

GHBDH can also convert glyoxylate to glycolate [19, 21]; the latter can be considered a major hypoxic metabolite accumulating under low oxygen stress [34]. It can be also considered an end product of fermentation originating from the nonphotorespiratory pathways of serine synthesis that represent alternative pathways of glycolysis [35]. Serine formed in these pathways is metabolized to glycine, glyoxylate, and then glycolate. To measure the glyoxylate reductase activity of GHBDH, SSA in the assay medium should be replaced by 5 mM glyoxylate [36].

4 Notes

1. Assays of the enzymes involved in glutamate and GABA metabolism are important for studying changes in nitrogen metabolism in plants in stress conditions which reflect accumulation of stress metabolites such as GABA and GHB and provide important information on rearrangement of the metabolic pathways of nitrogen metabolism to cope with the changes in pH and redox level in stressed cells. The assessment of enzyme activities described in this study represents valuable characterization of metabolic shifts and flexibility in response to environmental changes.
2. The application of biochemical methods for the study of glutamate and GABA conversions has certain limitations which consist in the relative instability of the enzymes, in their dependence on experimental conditions such as pH, temperature, and substrate concentration. The particular difficulty in the methods of measurement of GAD and GABA-T consists in the necessity of using auxiliary enzymes which may increase the inaccuracies in measurements. This means that in more advanced studies the traditional methods of activity measurement are usually complemented by other methods including Western blot analysis with monoclonal antibodies, molecular

genetic methods such as quantitative PCR with gene specific primers, and metabolomics profiling with NMR spectroscopy and mass spectrometry. However, despite the obvious advantages of these modern methods, they usually remain quite costly, time-consuming and do not always provide direct information on functional operation of enzymes; therefore, the use of the biochemical methods of analysis of enzyme activities remains feasible and they are often used for the diagnostic detection of stress reactions in plants.

Acknowledgments

Funding: This work was supported by the grant 19-14-00150 of the Russian Science Foundation.

References

1. Mano J (2012) Reactive carbonyl species: their production from lipid peroxides, action in environmental stress, and the detoxification mechanism. Plant Physiol Biochem 59:90–97
2. Kinnersley AM, Turano FJ (2000) γ-Aminobutyric acid (GABA) and plant responses to stress. Crit Rev Plant Sci 19:479–509
3. Popov VN, Eprintsev AT, Fedorin DN, Fomenko OY, Igamberdiev AU (2007) Role of transamination in the mobilization of respiratory substrates in germinating seeds of castor oil plants. Appl Biochem Microbiol 43:341–346
4. Fait A, Fromm H, Walter D, Galili G, Fernie AR (2008) Highway or byway: the metabolic role of the GABA shunt in plants. Trends Plant Sci 13:14–19
5. Khan MI, Nazir F, Asgher M, Per TS, Khan NA (2015) Selenium and sulfur influence ethylene formation and alleviate cadmium-induced oxidative stress by improving proline and glutathione production in wheat. J Plant Physiol 173:9–18
6. Yang H, Sun M, Lin S, Guo Y, Yang Y, Zhang T, Zhang J (2017) Transcriptome analysis of *Crossostephium chinensis* provides insight into the molecular basis of salinity stress responses. PLoS One 12:e0187124
7. Bown AW, Shelp BJ (1997) The metabolism and functions ofγ-aminobutyric acid. Plant Physiol 115:1–5
8. Michaeli S, Fait A, Lagor K, Nunes-Nesi A, Grillich N, Yellin A, Bar D, Khan M, Fernie AR, Turano FJ, Fromm H (2011) A mitochondrial GABA permease connects the GABA shunt and the TCA cycle, and is essential for normal carbon metabolism. Plant J 67:485–498
9. Renault H, El Amrani A, Berger A, Mouille G, Soubigou-Taconnat L, Bouchereau A, Deleu C (2013) γ-Aminobutyric acid transaminase deficiency impairs central carbon metabolism and leads to cell wall defects during salt stress in Arabidopsis roots. Plant Cell Environ 36:1009–1018
10. Steward FC, Thompson JF, Dent CE (1949) γ-Aminobutyric acid: a constituent of the potato tuber? Science 110:439–440
11. Corrales AR, Carrillo L, Lasierra P, Nebauer SG, Dominguez-Figueroa J, Renau-Morata B, Pollmann S, Granell A, Molina RV, Vicente-Carbajosa J, Medina J (2017) Multifaceted role of cycling DOF factor 3 (CDF3) in the regulation of flowering time and abiotic stress responses in Arabidopsis. Plant Cell Environ 40:748–764
12. Signorelli S, Dans PD, Coitiño EL, Borsani O, Monza J (2015) Connecting proline and γ-aminobutyric acid in stressed plants through non-enzymatic reactions. PLoS One 10: e0115349
13. Bowsher CG, Lacey AE, Hanke GT, Clarkson DT, Saker LR, Stulen I, Emes MJ (2007) The effect of Glc6P uptake and its subsequent oxidation within pea root plastids on nitrite reduction and glutamate synthesis. J Exp Bot 58:1109–1118
14. Breitkreuz KE, Allan WL, Van Cauwenberghe OR, Jakobs C, Talibi D, Andre B, Shelp BJ (2003) A novel gamma-hydroxybutyrate dehydrogenase: identification and expression of an

Arabidopsis cDNA and potential role under oxygen deficiency. J Biol Chem 278:41552–41556

15. Busch KB, Fromm H (1999) Plant succinic semialdehyde dehydrogenase. Cloning, purification, localization in mitochondria, and regulation by adenine nucleotides. Plant Physiol 121:589–598
16. Busch KB, Piehler J, Fromm H (2000) Plant succinic semialdehyde dehydrogenase: dissection of nucleotide binding by surface plasmon resonance and fluorescence spectroscopy. Biochemistry 39:10110–10117
17. Andriamampandry C, Siffert J-C, Schmitt M, Garnier J-M, Staub A, Muller C, Gobaille S, Mark J, Maitre M (1998) Cloning of a rat brain succinic semialdehyde reductase involved in the synthesis of the neuromodulator gamma-hydroxybutyrate. Biochem J 334:43–50
18. Schaller M, Schaffhauser M, Sans N, Wermuth B (1999) Cloning and expression of succinic semialdehyde reductase from human brain. Identity with aflatoxin B1 aldehyde reductase. Eur J Biochem 265:1056–1060
19. Allan WL, Simpson JP, Clark SM, Shelp BJ (2008) Gamma-hydroxybutyrate accumulation in Arabidopsis and tobacco plants is a general response to abiotic stress: putative regulation by redox balance and glyoxylate reductase isoforms. J Exp Bot 59:2555–2564
20. Drew MC (1997) Oxygen deficiency and root metabolism: injury and acclimation under hypoxia and anoxia. Annu Rev Plant Physiol Plant Mol Biol 48:223–250
21. Allan WL, Clark SM, Hoover GJ, Shelp BJ (2009) Role of plant glyoxylate reductases during stress: a hypothesis. Biochem J 423:15–22
22. Tsai K-J, Lin C-Y, Ting C-Y, Shih M-C (2016) Ethylene-regulated glutamate dehydrogenase fine-tunes metabolism during anoxia-reoxygenation. Plant Physiol 172:1548–1562
23. Fontaine JX, Tercé-Laforgue T, Armengaud P, Clément G, Renou JP, Pelletier S, Catterou M, Azzopardi M, Gibon Y, Lea PJ, Hirel B, Dubois F (2012) Characterization of a NADH-dependent glutamate dehydrogenase mutant of Arabidopsis demonstrates the key role of this enzyme in root carbon and nitrogen metabolism. Plant Cell 24:4044–4065
24. Melo-Oliveira R, Oliveira IC, Coruzzi GM (1996) Arabidopsis mutant analysis and gene regulation define a nonredundant role for glutamate dehydrogenase in nitrogen assimilation. Proc Natl Acad Sci U S A 93:4718–4723
25. Loyola-Vargas VM, de Jimenez ES (1984) Differential role of glutamate dehydrogenase in nitrogen metabolism of maize tissues. Plant Physiol 76:536–540
26. Glevarec G, Bouton S, Jaspard E, Riou MT, Cliquet JB, Suzuki A, Limami AM (2004) Respective roles of the glutamine synthetase/glutamate synthase cycle and glutamate dehydrogenase in ammonium and amino acid metabolism during germination and post-germinative growth in the model legume *Medicago truncatula*. Planta 219:286–297
27. Bhatnagar P, Glasheen BM, Bains SK, Long SL, Minocha R, Walter C, Minocha SC (2001) Transgenic manipulation of the metabolism of polyamines in poplar cells. Plant Physiol 125:2139–2153
28. Hiraga K, Ueno Y, Oda K (2008) Glutamate decarboxylase from *Lactobacillus brevis*: activation by ammonium sulfate. Biosci Biotechnol Biochem 72:1299–1306
29. Huang Y, Su L, Wu J (2016) Pyridoxine supplementation improves the activity of recombinant glutamate decarboxylase and the enzymatic production of gamma-aminobutyric acid. PLoS One 11:e0157466
30. Park JY, Jeong SJ, Kim JH (2014) Characterization of a glutamate decarboxylase (GAD) gene from *Lactobacillus zymae*. Biotechnol Lett 36:1791–1799
31. Hampe CS, Hammerle LP, Falorni A, Robertson J, Lernmark A (2001) Site-directed mutagenesis of K396R of the 65 kDa glutamic acid decarboxylase active site obliterates enzyme activity but not antibody binding. FEBS Lett 488:185–189
32. Nonaka S, Someya T, Zhou S, Takayama M, Nakamura K, Ezura H (2017) An *Agrobacterium tumefaciens* strain with gamma-aminobutyric acid transaminase activity shows an enhanced genetic transformation ability in plants. Sci Rep 7:42649
33. Saito N, Robert M, Kochi H, Matsuo G, Kakazu Y, Soga T, Tomita M (2009) Metabolite profiling reveals YihU as a novel hydroxybutyrate dehydrogenase for alternative succinic semialdehyde metabolism in *Escherichia coli*. J Biol Chem 284:16442–16451
34. Narsai R, Howell KA, Carroll A, Ivanova A, Millar AH, Whelan J (2009) Defining core metabolic and transcriptomic responses to oxygen availability in rice embryos and young seedlings. Plant Physiol 151:306–322
35. Igamberdiev AU, Kleczkowski LA (2018) The glycerate and phosphorylated pathways of serine synthesis in plants: the branches of plant glycolysis linking carbon and nitrogen metabolism. Front Plant Sci 9:318
36. Kleczkowski LA, Edwards GE (1989) Identification of hydroxypyruvate and glyoxylate reductases in maize leaves. Plant Physiol 91:278–286

Chapter 8

Using Different Forms of Nitrogen to Study Hypersensitive Response Elicited by Avirulent *Pseudomonas syringae*

Namrata Singh, Pooja Singh, Pradeep Kumar Pathak, and Kapuganti Jagadis Gupta

Abstract

Nitrate, ammonium, or a combination of both is the form of N available for nitrogen assimilation from soil by the plants. Nitrogen is an important and integral part of amino acids, nucleotides, and defense molecules. Hence it is very important to study the role of nitrate and ammonium nutrition in plant defense via hypersensitive response (HR). Shifting plants from ammonium nitrate Hoagland solution to nitrate Hoagland nutrition slightly enhances root length and leaf area. HR phenotype is different in nitrate and ammonium grown plants when challenged with avirulent *Pseudomonas syringae* DC3000 avrRpm1. HR is also associated with increased production of reactive oxygen species (ROS) and nitric oxide (NO). Hence to understand HR development it is essential to measure HR lesions, cell death, ROS, NO, and bacterial growth. Here we provide a stepwise protocol of various parameters to study HR in *Arabidopsis* in response to nitrate and ammonium nutrition.

Key words Nitrate, Ammonium, Nitric oxide, Reactive oxygen species, Hypersensitive response, *Pseudomonas syringae*

1 Introduction

Nitrogen is an essential element which is required for the plant growth, development and resistance against wide range of biotic and abiotic stresses. Nitrogen also effects cell size, growth, and cell wall composition [1]. Plants obtain nitrogen mainly from two different sources (1) atmospheric nitrogen via symbiotic nitrogen fixation and (2) from soil in various forms, mainly nitrate, ammonium, or a combination of both. The availability of different nitrogen forms in soil depends on various factors that include formation of inorganic nitrogen from organic nitrogen and further nitrification and denitrification processes [2]. These processes are influenced by soil conditions, crop cultivar, soil pH, oxygen availability, availability of minerals such as NaCl and conditions such as temperature [3–5]. Nitrate is first converted to ammonia then

Kapuganti Jagadis Gupta (ed.), *Nitrogen Metabolism in Plants: Methods and Protocols*, Methods in Molecular Biology, vol. 2057, https://doi.org/10.1007/978-1-4939-9790-9_8,

incorporated to amino acids such as aspargine, aspartate, glutamine, and glutamate [6]. Plant can take up ammonium directly.

The ratio of nitrate (NO_3^-) or ammonium (NH_4^+) significantly affects growth and also plant defense against biotic and abiotic stress [7–10]. It has been shown that plant N-status affects susceptibility of plants toward fungal and bacterial pathogens [8, 9, 11]. Nitrogen fertilization affects the outcome of plant pathogen interaction. For instance, application of N-fertilizer in the form of NO_3^- generally enhances HR-linked resistance; a phenomenon that was associated with increased generation of nitric oxide (NO), ROS, biosynthesis of the key defense signal molecule salicylic acid (SA) and polyamines accumulation [8] whereas, ammonium nutrition increases susceptibility via production of apoplastic sugar, γ aminobutyric acid (GABA) a glutamate-derived metabolite and very high levels of ROS generation. Under nitrate nutrition nitric oxide a free radical molecule production is predominant which plays a role in increased resistance via production of peroxynitrite ($ONOO^-$) during HR. Nitric oxide is produced by various enzymes [12]. Growing plants on nitrate nutrition leads to activation of nitrate reductase via production of intermediate nitrite [13]. Previous it was shown that nitrate and ammonium nutrition also influence reactive oxygen species which are important for development of HR [14]. Their ratios play a role in determining HR [8] and production of NO [15].

Hence in order to understand plant resistance against pathogens in response to various N regimes, there is a need to analyze various parameters such as observation of phenotype, cell death, analysis of bacterial growth, and determination of ROS and NO [16]. Hence here we describe in detail a protocol to analyze the plant resistance against bacterial pathogen avr *Pseudomonas syringae* DC3000 in *Arabidopsis* under nitrate, ammonium, or a combination of both (control) medium.

2 Materials

2.1 Plant Material

Arabidopsis thaliana wild type (Col-0) plants.

2.2 Chemicals Required

1. Sodium hypochlorite (4%).
2. 10 mM $MgCl_2$.
3. 100% ethanol.
4. Hydroponics Medium (Hoagland's medium):

   Control medium: Macronutrients: 1 mM NH_4Cl, 1 mM KNO_3 1 mM $CaCl_2$, 1 mM $MgSO_4$, 1 mM·K_2HPO_4, 1 mM KH_2PO_4 and micronutrients: 15 μM H_3BO_3, 3 μM·$MnCl_2 \cdot 4H_2O$, 0.25 μM $ZnSO_4 \cdot 7H_2O$, 0.1 μM $CuSO_4 \cdot 5H_2O$, 0.04 μM Na_2MoO_4, 25 μm NaFe–EDTA.

Nitrate medium: Hoagland medium containing only 3 mM KNO_3 as N source.

Ammonium medium: Hoagland medium containing only 1 mM NH_4Cl as N source.

5. 3,3′-Diaminobenzidine (HiMedia).
6. Tween 20 liquid (molecular biology grade).
7. Sodium phosphate (Na_2HPO_4).
8. HCl.
9. Acetic acid.
10. Glycerol.
11. Nitrotetrazolium blue chloride (NBT).
12. 50 mM sodium phosphate buffer (pH 7.5).
13. Trypan blue.
14. Lactic acid 85% (w/w).
15. Phenol solution (TE buffer equilibrated, pH 7.5–8.0).
16. Glycerol ≥99%.
17. 100 mm HEPES buffer (PH 7.2).
18. 4-amino-5-methylamino-2′,7′-difluorofluorescein (DAF-FM).
19. Carboxy-PTIO potassium salt.
20. $MgSO_4$.
21. Dipotassium hydrogen orthophosphate (K_2HPO_4).
22. Peptone.
23. Kings medium B Base (HiMedia).

2.3 Equipment Required

1. Plant growth facility (culture room): 8 h light and 16 h dark period, day/night temperature of 21/19 °C and relative humidity 55–60%.
2. Cold room.
3. Fluorescence microscope (Nikon ECLIPSE-80i, Japan).
4. Nikon stereozoom AZ 100 microscope.
5. Shaker.
6. Autoclave.
7. Hot air oven (60 °C).
8. Weighing balance.
9. Magnetic stirrer.
10. PH meter.
11. Biological safety cabinet.
12. Nikon COOLPIX camera.

2.4 General Items Required

1. Square petri dishes disposable (120 × 120 mm).
2. Round petri dishes disposable (90 mm).
3. 6-well microtiter plate.
4. Glass slides.
5. Coverslips.
6. Dropper.
7. Scissors.
8. Plastic cups disposable.
9. Razor blade.
10. Forceps.
11. Brush (0 size).
12. Glass marker pens, stickers, tags.
13. Sterile MQ water.
14. Tooth picks.
15. Whatman paper.
16. Soilrite and agro peat (1:1).
17. Measuring cylinder.
18. Aluminum foil.
19. Tips.
20. Conical flasks.
21. 50 ml Falcon tube.
22. Para film.
23. 1 ml needleless syringe.
24. Cork borer.
25. Scale.

3 Methodology

3.1 Surface Sterilization of Seeds and Plant Growth

1. Use 2% sodium hypochlorite solution for surface sterilization.
2. Add minimum quantity of sterilized MQ into the 1.5 ml microcentrifuge tube containing the seeds (200–300 seeds).
3. Soak it for 30 min at 4 °C.
4. Remove the MQ from the tube and add 1 ml of 2% sodium hypochlorite solution into Eppendorf tube containing the seeds. Incubate for 7–8 min, then wash with autoclaved distilled water for 6–7 times.

3.2 Plant Growth on ½ MS

1. Prepare ½ strength MS medium containing 2.212 g MS, 15 g sucrose, and 8 g of agar in 1000 ml distilled water.

2. Autoclave at 121 °C for 15 min.
3. Pour the media into square petri plates.
4. Once the media solidifies, place the seeds using pipette on ½ MS plates and store at 4 °C for 3 days for stratification.
5. After stratification transfer the plates into culture room under 8 h light and 16 h dark period (SD) maintain day/night temperature of 21/19 °C and relative humidity 55–60%.
6. Once the seedlings are 10 days old, transfer into individual pots containing Soilrite and agro peat (1:1).

3.3 Preparation of Hydroponics Medium

1. Prepare nitrate, ammonium, and control medium in MQ water and set the pH to 6.3.
2. Once the plants are 3 weeks old transfer them into plastic cups (hydroponics) (Fig. 1a) containing nitrate, ammonium, and control medium and allow the plants to adapt.
3. Keep the hydroponics cups into culture room under short day condition (8 h light and 16 h dark period).
4. Consecutively change the nutrient solutions in every 2–3 days.
5. Allow the plants to grow for 10 days. Observe the root phenotype (Fig. 1b).

3.4 Root Length and Leaf Area Measurement After Nitrate, Ammonium, and Control Treatment

1. Place the cup on filter paper.
2. Carefully hold the plant containing lid.
3. Measure the root length (Fig. 1b).
4. Immediately place the plant into the cup along with lid.
5. For measuring leaf area place the scale and measure the length and width and calculate the area (Fig. 2).

3.5 Growth of Bacteria

3.5.1 KB Broth Preparation

1. Mix 2 g of peptone, 0.15 g of K_2HPO_4, and 0.15 g of $MgSO_4 \cdot 7H_2O$ into 100 ml of distilled water.
2. Autoclave at 121 °C for 15 min for sterilization.

3.5.2 Primary Culture Preparation

1. Inoculate 20 μl glycerol stock of *Pseudomonas syringae* DC 3000 AvrRpm1 into 5 ml of KB broth containing rifampicin (50 mg/ml), then incubate at 28 °C and 200 rpm for 48 h.

3.5.3 Secondary Culture Preparation

1. Inoculate 1 ml of primary culture into 100 ml KB broth and incubate at 28 °C and 200 rpm for 12 h.

3.5.4 Pathogen Working Stock Preparation

1. Precool the centrifuge to 4 °C.
2. Centrifuge the bacterial culture at 2200 × *g* for 10 min.

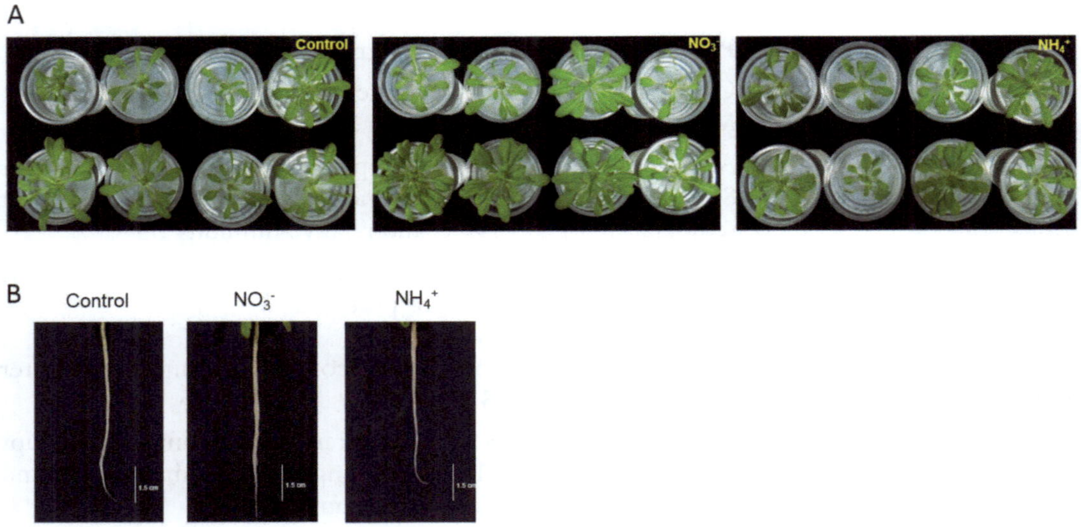

Fig. 1 (**a**) Phenotype of 4-week-old *Arabidopsis* plants growing on hydroponics solution. Plants were transferred to hydroponics at 3-week-old stage from soil. (**b**) Roots of hydroponically grown plants

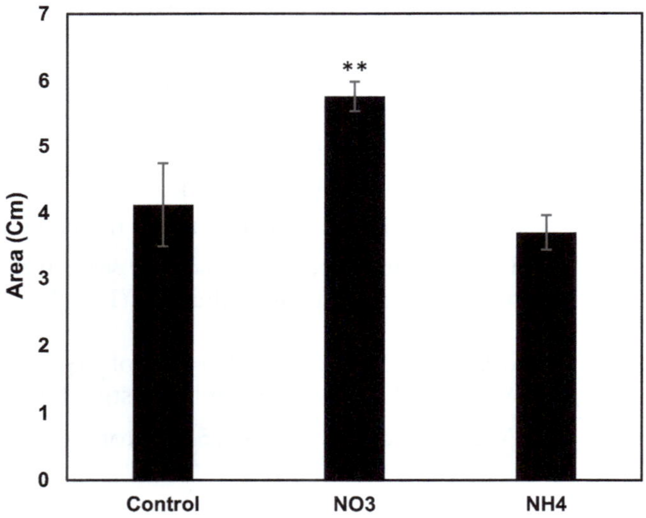

Fig. 2 Leaf area of hydroponically shifted plants on control, NO_3, or NH_4. At 3 week old stage, plants were shifted from soil to hydroponics and observed phenotype on tenth day. Error bars represent $n = 4 \pm$ SD. Statistical significance is determined by *t*-test with respect to control treated plants. ($**$ sign represent *P* values <0.02)

3. Discard the media into a flask and dissolve the pellet in 1 ml of autoclaved 10 mM $MgCl_2$.
4. Resuspend the pellet in 10 mM $MgCl_2$ until OD reaches to 0.2, which correspond to approximately 10^8 CFU/ml.

Fig. 3 Hypersensitive response phenotype in control and NO_3 or NH_4 treated plants challenged with avirulent *Pseudomonas syringae* DC3000 after 24 h post infiltration. Image is representative of three biological replicates

3.5.5 Pathogen Infiltration to the Plants

1. Infiltrate *Arabidopsis* plants with avr *Pseudomonas syringae* DC 3000 by using 1 ml needleless syringe (*see* **Note 1**).
2. After 24 h of infiltration take pictures (Fig. 3) to observe hypersensitive response (HR) phenotype.

3.6 Trypan Blue Staining

3.6.1 Staining Solution Preparation

1. Mix 10 ml lactic acid (85% w/w), 10 ml phenol (TE buffer equilibrated, pH 7.5–8.0), 10 ml glycerol ($\geq$ 99%), 10 ml of autoclaved distilled water and 40 mg of trypan blue stain in a flask and make the final concentration of 1 mg/ml solution (*see* **Note 2**).

3.6.2 Staining Procedure

1. After 24 h post infiltration, harvest the leaves by using scissors.
2. Place the leaves into plates containing trypan blue staining solution (1 mg/ml) and incubate them for 8 h.
3. Remove the staining solution and refill the plates with absolute ethanol (100% ethanol) to remove the chlorophyll.
4. Incubate tissue in ethanol for at least 12 h and replace the ethanol with fresh solution for every 3–4 h until the chlorophyll is removed.
5. After chlorophyll removal, replace the ethanol solution with 50% glycerol.
6. Capture picture of the leaf (Fig. 4).

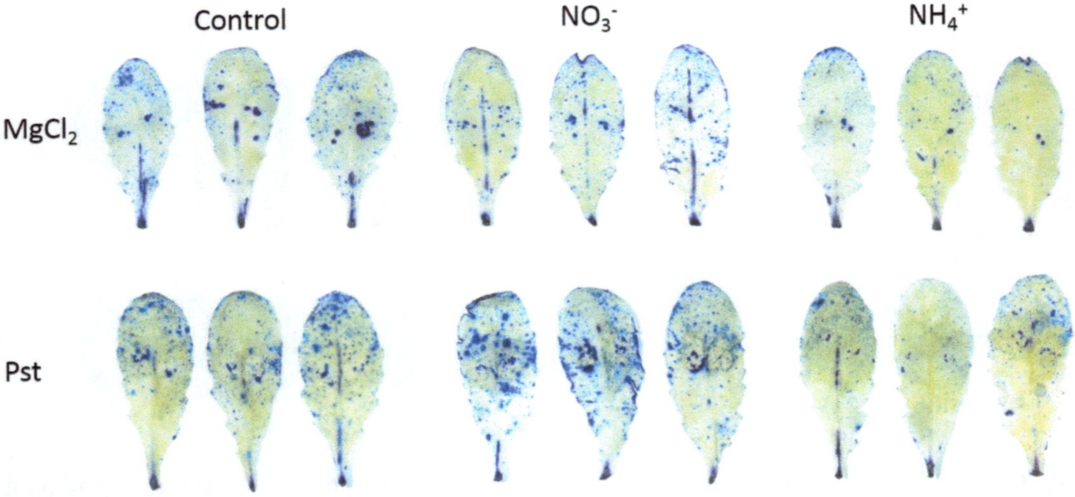

Fig. 4 Trypan blue staining in control, NO_3 or NH_4 treated plants challenged with avirulent *P. syringae* DC3000 after 24 h post infiltration. Image is a representative of three biological replicates

3.7 Bacterial Growth

3.7.1 KBA Plates Preparation

1. Dissolve 42.23 g of King's medium B Base in 1000 ml distilled water.
2. Mix completely by using magnetic stirrer.
3. Autoclave at 121 °C for 15 min.
4. Aseptically pour into sterile round Petri plates inside the laminar flow hood.

3.7.2 CFU Count

1. Cut the leaf discs with cork borer from infected leaves and sterilize with 0.2% H_2O_2 for 8 min.
2. Remove the leaf disc with forceps and place in fresh microcentrifuge tube containing 1 ml of 10 mM $MgCl_2$.
3. Crush the leaf discs by using micro pestle (Fig. 5a).
4. Serially dilute until 10^7 dilution factor.
5. Spread 100 μl of the diluted bacterial on KBA plates containing Rifampicin (10 mg/ml) and keep at 28 °C for 48 h (Fig. 5b).
6. Count bacteria manually after 48 h of incubation (Fig. 5c).

3.8 DAB Staining

3.8.1 Staining Solution Preparation

1. Add 50 mg of DAB and 45 ml of sterilized distilled water in a 100 ml conical flask to prepare final concentration of DAB solution (i.e., 1 mg/ml) (*see* **Note 3**).
2. Add magnetic stirrer and adjust the pH of staining solution to 3.0 with 1 N HCl to dissolve DAB.
3. Add 25 μl of tween 20 (0.05% v/v) and 2.5 ml 200 mM Na_2HPO_4 into the solution.

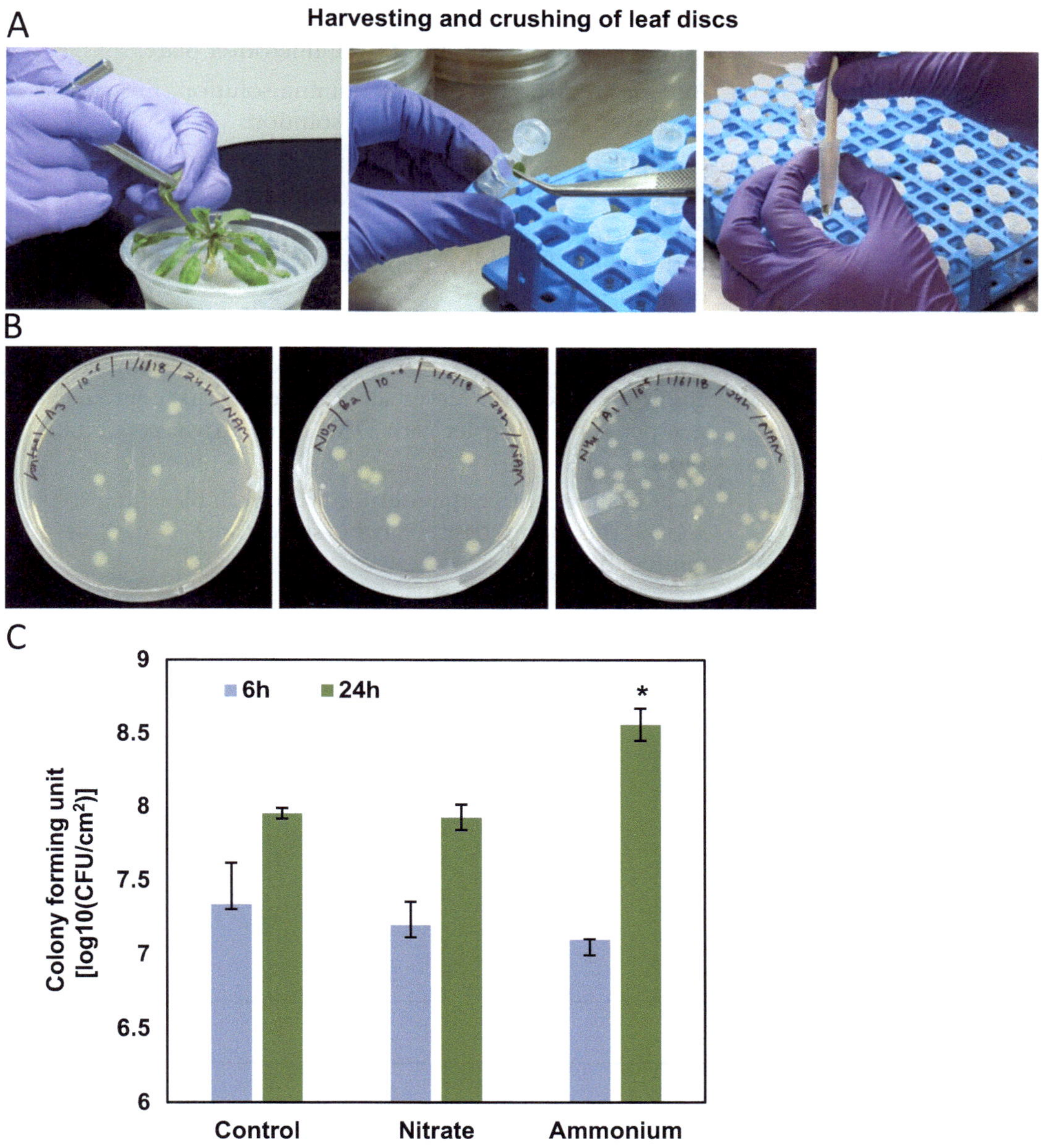

Fig. 5 Bacterial levels in control, NO_3 or NH_4 treated WT plants inoculated with avr *P. syringae* DC3000 was monitored at 6 and 24 hpi. Bars represent the SD values. Statistical significance is determined by *t*-test with respect to control treated plants. (Asterisk (∗) sign represents *P* values <0.05)

4. Adjust the final volume into 50 ml. This will generate 10 mM Na_2HPO_4 DAB solution which will pull the pH back up again.

3.8.2 Staining Procedure

1. Place the DAB solution in 6-well microtiter plate.
2. Harvest pathogen infiltrated leaves by using scissors. Sample at least 3 leaves per plant from four independent biological replicates.

3. Cut the leaves at the desired time (In this case 24 h) point by using scissors and put it in 6-well microtiter plate.
4. Add enough volume of DAB staining solution. Ensure that all the leaves are submerged in the solution.
5. Incubate the leaves in staining solution for 5–6 h. Plates should be covered with aluminum foil (*see* **Note 4**).
6. After incubation remove the DAB staining solution and replace with the Bleaching solution (Ethanol: acetic acid: glycerol = 3:1:1).
7. Place the 6-well plate in a boiling water bath (~90–95 °C temperature) for 10–15 min. This process will remove all the chlorophyll from the leaves, and only leaves that contain hydrogen peroxide will react with DAB form brown precipitate (*see* **Note 5**).
8. After boiling replace the solution with fresh bleaching solution and at this point tissue can be store for 3–4 days at 4 °C temperature.
9. Capture the image with camera or stereomicroscope (Fig. 6).

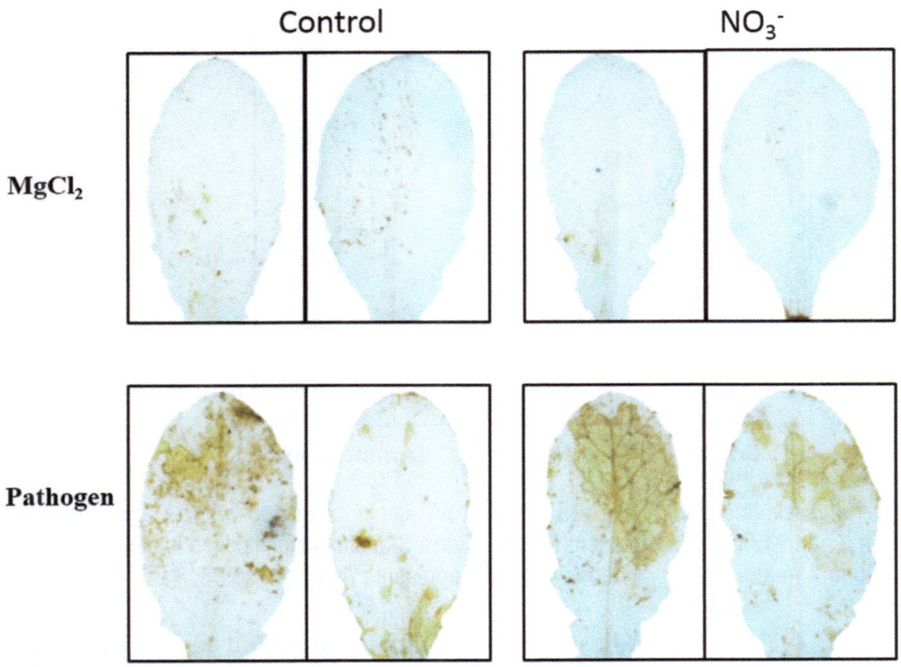

Fig. 6 Picture showing DAB staining of *P. syringae* DC3000 treated and mock infiltrated leaves after 24 h of infiltration

3.9 NBT Staining

3.9.1 Staining Solution Preparation

1. Dissolve 0.1 g NBT in 50 mM sodium phosphate buffer and make the volume 50 ml to get 0.2 mg/ml solution.
2. Cover the flask with aluminum foil because NBT is light sensitive.
3. Mix the solution by using magnetic stirrer.

3.9.2 Staining Procedure

1. Place the NBT staining solution in 6-well microtiter plate.
2. Harvest the leaves of desired time point by using scissors and put into 6-well plate containing NBT solution. Insure that leaves are immerged into the solution.
3. Cove the plate with aluminum foil and incubate for 5–6 h on a shaker.
4. Remove the staining solution after 5–6 h and refill with absolute ethanol to remove chlorophyll from the leaves.
5. Put the 6-well plate into boiling water bath (~90–95 °C temperature) for 10–15 min depending upon the sample. After boiling all the chlorophyll will remove and only dark blue stain of formazan compound will remain.
6. Replace the solution with fresh 50% ethanol solution. It can be stored at 4 °C temperature for 3–4 days.
7. Capture the image using camera or stereomicroscope (Fig. 7).

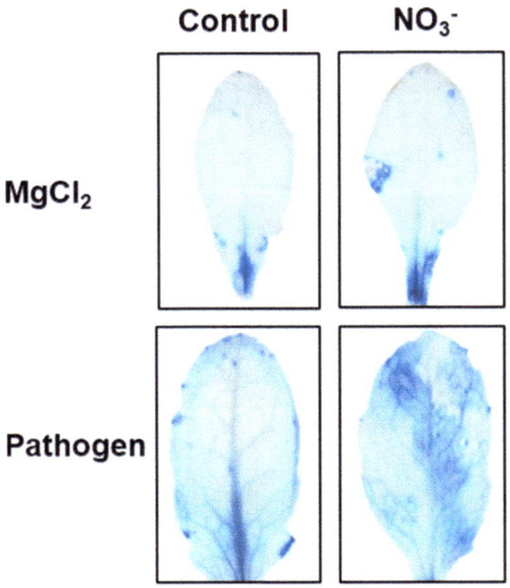

Fig. 7 Picture showing NBT staining of leaves challenged with avirulent *P. syringae* DC3000 and mock infiltrated leaves after 24 h of infiltration

3.10 Nitric Oxide (NO) Estimation by Using DAF-FM Dye

1. Infiltrate *Arabidopsis* leaves with 10 µm DAF-FM dye prepared in 100 mM HEPES buffer (pH 7.2) by using 1 ml needleless syringe.
2. Incubate the plants in dark for 30 min. For control infiltrate 200 µm cPTIO along with 10 µM DAF-FM dye into the treated plants under same set of conditions.
3. After incubation remove the epidermis from abaxial side of the leaves by using fine forceps.
4. Cut leaves into small segments around the infiltrated region.
5. Wash the leaf segments 2–3 times by using 100 mm HEPES buffer.
6. Put the segments onto the slide and place cover slip on it.
7. Visualize the sections under fluorescence microscope with 495 nm excitation and 515 nm emission wavelength and capture image (Fig. 8).

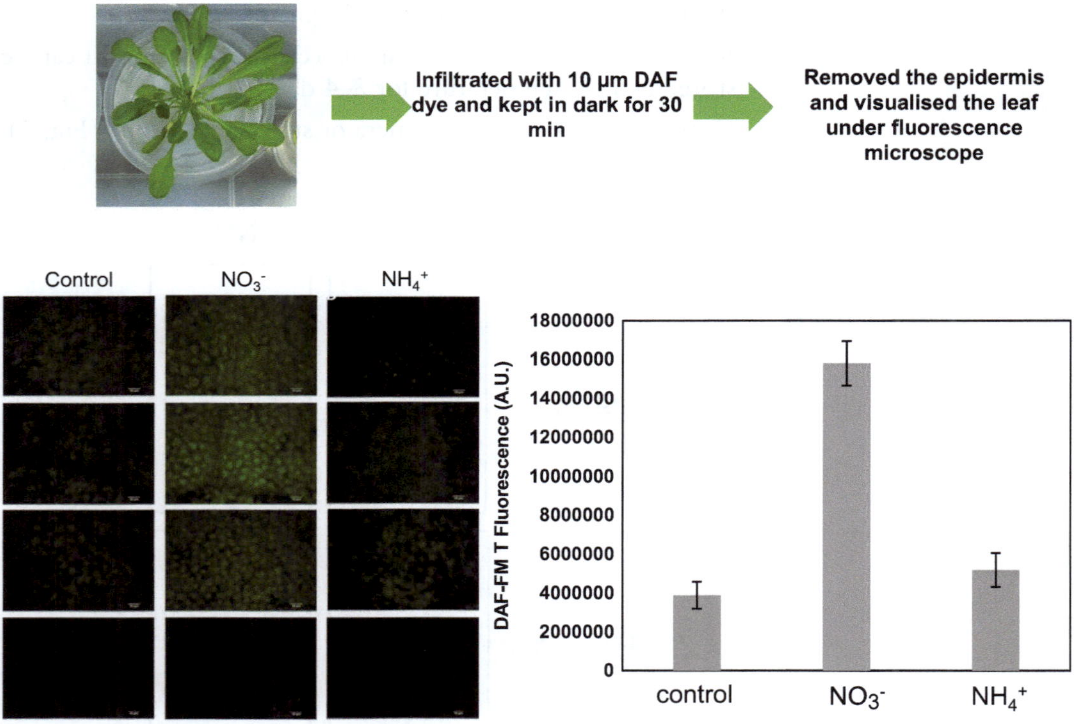

Fig. 8 Localization of NO production in *Arabidopsis* plants, grown on Hoagland's media supplemented with different N modified nutrient solutions

4 Notes

1. Infiltrate slowly to avoid wounding.
2. All the staining solutions should be freshly prepared.
3. While performing DAB staining always take laboratory precautions.
4. Staining solutions are light sensitive so that it has to be covered with aluminum foil.
5. For chlorophyll removal, boiling temperature should be depends on the type of tissue.

Acknowledgments

NS is a recipient of JRF and SRF from CSIR, India. PS and PKP are recipients of UGC JRF and SRF. This work is supported by SERB Early Carrier Research Award to KJG by SERB, Govt. of India.

References

1. Huber DM, Watson RD (1974) Nitrogen form and plant disease. Annu Rev Phytopathol 12:139–165
2. Giles ME, Morley NJ, Baggs EM, Daniell TJ (2012) Soil nitrate reducing processes–drivers, mechanisms for spatial variation, and significance for nitrous oxide production. Front Microbiol 3:407
3. Simek M, Cooper JE (2002) The influence of soil pH on denitrification: progress towards the understanding of this interaction over the last 50 years. Eur J Soil Sci 53:345–354
4. Gebauer RL, Tenhunen JD, Reynolds JF (1996) Soil aeration in relation to soil physical properties, nitrogen availability, and root characteristics within an arctic watershed. Plant Soil 178:37–48
5. Veresoglou SD, Voulgari OK, Sen R, Mamolos AP, Veresoglou DS (2011) Effects of nitrogen and phosphorus fertilization on soil pH-plant productivity relationships in upland grasslands of northern Greece. Pedosphere 21:750–752
6. Lam HM, Coschigano KT, Oliveira IC, Melo-Oliveira R, Coruzzi GM (1996) The molecular-genetics of nitrogen assimilation into amino acids in higher plants. Annu Rev Plant Biol 47:569–593
7. Wang P, Wang Z, Pan Q, Sun X, Chen H, Chen F, Mi G (2019) Growth of maize on mixed nitrate and ammonium promotes auxin synthesis and biomass accumulation. J Exp Bot. https://doi.org/10.1093/jxb/erz047
8. Gupta KJ, Brotman Y, Segu S, Zeier T, Zeier J, Persijn ST, Cristescu SM, Harren FJM, Bauwe H, Fernie AR (2013) The form of nitrogen nutrition affects resistance against *Pseudomonas syringae pv. phaseolicola* in tobacco. J Exp Bot 64:553–568
9. Snoeijers SS, Pérez-García A, Joosten MH, De Wit PJ (2000) The effect of nitrogen on disease development and gene expression in bacterial and fungal plant pathogens. Eur J Plant Pathol 106:493–506
10. Wany A, Gupta KJ (2018) Reactive oxygen species, nitric oxide production and antioxidant gene expression during development of aerenchyma formation in wheat. Plant Signal Behav 13:3002–3017
11. Lopez-Berges MS, Rispail N, Prados-Rosales RC, Di Pietro A (2010) A nitrogen response pathway regulates virulence in plant pathogenic fungi: role of TOR and the bZIP protein Mea B. Plant Signal Behav 5:1623–1625
12. Gupta KJ, Fernie AR, Kaiser WM, van Dongen JT (2011) On the origins of nitric oxide. Trends Plant Sci 16:160–168
13. Planchet E, Gupta KJ, Sonoda M, Kaiser WM (2005) Nitric oxide emission from tobacco leaves and cell suspensions: rate limiting factors and evidence for the involvement of mitochondrial electron transport. Plant J 41:732–743

14. Podgorska A, Ostaszewska M, Gardeström P, Rasmusson AG, SZAL B (2015) In comparison with nitrate nutrition, ammonium nutrition increases growth of the frostbite1 Arabidopsis mutant. Plant Cell Environ 38:224–237

15. Delledonne M, Xia Y, Dixon RA, Lamb C (1998) Nitric oxide functions as a signal in plant disease resistance. Nature 394:585

16. Wany A, Gupta A, Kumari A, Mishra S, Singh N, Pandey S, Vanvari R, Igamberdiev A, Ferine A, Gupta KJ (2018) Nitrate nutrition influences multiple factors in order to increase energy efficiency under hypoxia in Arabidopsis. Ann Bot. https://doi.org/10.1093/aob/mcy202

Chapter 9

Using Foldscope to Monitor Superoxide Production and Cell Death During Pathogen Infection in Arabidopsis Under Different Nitrogen Regimes

Reena Arora, Pooja Singh, Aprajita Kumari, Pradeep Kumar Pathak, and Kapuganti Jagadis Gupta

Abstract

Nitrogen nutrition plays a role in plant growth development and resistance against biotic and abiotic stress. During pathogen infection various signal molecules such as reactive oxygen species, calcium, reactive nitrogen species, salicylic acid, and ethylene plays an important role. The form of nitrogen nutrition such as nitrate or ammonium plays a role in production of these molecules. Under nitrate nutrition NO is predominant. The produced NO plays a role in reacting with superoxide to generate peroxynitrite to induce cell death during hypersensitive response elicited by avirulent pathogens. Excess of ROS is also detrimental to plants and NO plays a role in regulating ROS. Hence it is important to observe superoxide production during infection. By using an avirulent *Pseudomonas syringae* and Arabidopsis differential N nutrition we show superoxide production in leaves using a paper microscope called Foldscope, which can be applied as a simple microscope to observe objects. The data also compared with root system infected with pathogenic *Fusarium oxysporum*. Taken together here we show that Foldscope is a cost-effective and powerful technique to visualize superoxide and cell death in plants during infection.

Key words Foldscope, Nitrate, Ammonium, Arabidopsis, *Pseudomonas syringae*, *Fusarium oxysporum*

1 Introduction

During the operation aerobic metabolism several forms of ROS are produced as superoxide ($O_2^{\bullet-}$), hydrogen peroxide (H_2O_2), hydroxyl radical ($OH^{\bullet}$), singlet oxygen ($^1O_2$) [1]. ROS are mainly produced by various cellular enzymes and compartments such as plasma membrane NADPH oxidase, peroxisomes, mitochondria, chloroplasts etc. ROS acts as signals when produced at physiological range and if produced in excess amount they can be very cytotoxic. For instance, low levels of ROS play role in modification

Reena Arora and Pooja Singh contributed equally to this work.

Kapuganti Jagadis Gupta (ed.), *Nitrogen Metabolism in Plants: Methods and Protocols*, Methods in Molecular Biology, vol. 2057, https://doi.org/10.1007/978-1-4939-9790-9_9, © Springer Science+Business Media, LLC, part of Springer Nature 2020 of cysteine residues of proteins leading to structural and conformational changes thus play a role in changing the activities of proteins. One of such prominent examples is oxidation of Cys thiolate anions (Cys–S$^-$) at to their sulfenic form (Cys–SOH) of proteins by (H_2O_2). This kind of modification can affect the function of proteins and thus play a role in signaling [2, 3]. Thus maintaining of basal levels of ROS is essential for redox based reactions and various processes such as regulation of cell cycle, stress recognition, stress acclimation, pathogen defense and programmed cell death [1]. If ROS are produced in larger amounts they cause DNA fragmentation, lipid peroxidation cytochrome c release, decreased ATP/ADP ratio, cytoplasmic acidosis, swelling of mitochondria, and loss of membrane integrity [4, 5]. These hallmarks of programmed cell death caused by increased reactive oxygen species. Hence plants have developed various antioxidant mechanisms to keep ROS low. For example, superoxide dismutases causes dismutation of O_2^- to H_2O_2, and catalase (CAT) dismutates H_2O_2 to oxygen and water, thereby reducing H_2O_2 concentration. Ascorbate peroxidase (APX) is an important enzyme involved in ascorbate glutathione cycle; it reduces H_2O_2 to water by using ascorbate (ASC) as an electron donor [6]. Copper amine oxidase, flavin polyamine oxidases, and oxalate oxidase also directly generate H_2O_2 [7]. It has shown that horseradish peroxidase can generate H_2O_2 during lignin biosynthesis [8]. Presence of high levels of H_2O_2 can produced toxic $^{\bullet}OH$ via interaction with transition metals [6]. H_2O_2 production also helpful to giving strength to cell walls during pathogen infections thus it aids in spread of pathogen from one cell to another cell. It also helps in activation of defense responses.

Transgenic manipulation of several antioxidants enzymes such as superoxide dismutase, catalase, and ascorbate peroxidase) exited differential cell death suggesting that redox balance between ROS and antioxidant plays a role in cell death.

In response to plant pathogen attack a fine balance between ROS and antioxidant defenses play important role. For instance, plants within few minutes of pathogen attack plants produce ROS [9]. Levine et al. [10] first demonstrated that hydrogen peroxide plays a role in development of cell death [4, 11, 12]. Increased antioxidant levels were observed in response to powdery mildew infection [13]. Suppression of APX was shown during HR development [14]. An increased induction of alternative oxidase was observed in response to TMV infection, elicitor treatments [15]. Increased induction of antioxidant defenses were observed in response to *Rhizoctonia solani* infection [16]. The induced AOX plays a role in regulation of ROS. It has been demonstrated that nitric oxide is an important molecule involved in induction of cell death via production of peroxynitrite through it reaction with superoxide [4]. The production of these free radicals may vary in nitrate or ammonium nutrition. Application of nitrate form of nutrition leads to increased resistance in tobacco plants challenge with avirulent and virulent *Pseudomonas* [17]. Nitrate nutrition generates NO via activation of nitrate reductase [17]. Nitrate reductase deficient *nia1,2* mutant was unable to produce NR and become susceptible to *Pseudomonas* infection. Application of nitrite elevated NO in *nia* mutant [18].

ROS measurement is required for assessment of plant resistance or susceptibility. Several methods are available for measurement of ROS such as Nitro blue tetrazolium (NBT) [19], Electron spin resonance spin-probe [20], and fluorescence techniques [5]. NBT is widely used to check superoxide during pathogen infection [21, 22].

Localized programed cell death is a feature of HR [23] in plants. HR can be detected and quantified by several methods such as measurement of released electrolytes into water in response to pathogen infiltration in the leaves [17, 24] and by using evans blue staining [25] and trypan blue staining method [26]. Trypan blue is a cell membrane impermeable azo dye and it can enter only in cells with ruptured membranes (dead cells). After entry in to the cell trypan blue binds to intracellular proteins therefor it allows direct detection of dead (blue) and live cells (unstained). Trypan blue was used to detect hypersensitive-response-associated necrosis in tobacco leaves [27], *Arabidopsis* leaves [28, 29] and other systems during plant pathogen interaction.

2 Material and Equipment

2.1 Bacterial Inoculation

1. King's Medium B base.
2. Weighing balance.
3. Autoclave.
4. Oak ridge tube.
5. Rifampicin (50 mg/mL).
6. *P. syringae* pv. tomato DC3000.
7. Incubator shaker (28 °C).
8. $MgCl_2$.
9. Conical flask.
10. Cotton plug.
11. Laminar hood.
12. Pipetman.
13. Culture vials.
14. Cuvette.
15. Spectrophotometer.

16. Pathogen infection room.
17. 1 mL syringe (Without Needle).
18. Nitrile gloves.

2.2 Fusarium oxysporum (f. sp. conglutinans) Inoculation

1. *Fusarium oxysporum* (f. sp. *conglutinans*) inoculum.
2. Potato dextrose agar (PDA).
3. 20 mm test tube.
4. Test tube stand.
5. Incubator (25 °C).
6. Blade.
7. Laminar hood.

2.3 Plant Growth

1. Murashige and Skoog medium (MS Medium).
2. Cold room.
3. Soilrite.
4. Agropete.
5. Plant growth chamber.
6. Seed sterilization solution (70% ethanol, 0.1% TritonX100).
7. 100% Ethanol.
8. Filter paper.
9. Micro centrifuge tube.
10. Hogland media (Macronutrients: 3 mM KNO_3, 3 mM NH_4Cl, 1 mM $CaCl_2$, 1 mM $MgSO_4$, 1 mM K_2HPO_4, 1 mM KH_2PO_4 and Micronutrients: 25 μM NaF-EDTA, 15 μM H_3BO_3, 3 μM $MnCl_2.4\ H_2O$, 0.25 μM $ZnSO_4.7\ H_2O$, 0.1 μM $CuSO_4.5H_2O$, and 0.04 μM Na_2MO_4.

2.4 Staining for NBT and Trypan Blue

1. Nitro blue tetrazolium (NBT).
2. Sodium phosphate buffer.
3. Trypan blue.
4. Lactic acid (85%).
5. Phenol.
6. Glycerol (99%).
7. Measuring cylinder.
8. Distilled water.
9. Beaker.
10. Brush.
11. Shaker.
12. Scissors.
13. Slide.

14. Coverslip.
15. Arabidopsis leaves (4 weeks old).
16. Arabidopsis root (15 days old).
17. Magnetic stirrer.
18. M*agnet beads.*
19. Aluminum foil.
20. Fold scope.

3 Methods

3.1 Growing Arabidopsis Plants

1. Sterilize *Arabidopsis* seeds with the seed sterilization solution (70% ethanol, 0.1% TritonX100) for 5 min. Short spin the microcentrifuge tube containing seed, discard the supernatant and add 1 mL 100% ethanol. Keep it for 5 min then transfer the seed on filter paper in laminar hood.
2. Keep the sterilized seeds on MS media plate.
3. Stratify the MS plates in cold room for 2 days in dark.
4. Transfer the plates to Arabidopsis growth chamber for 15 days at 22 °C in a photoperiod of 16 h of light/8 h of dark.
5. Transfer some seedlings (15 days old) to individual pots containing the Soilrite and agropeat (1:1 ratio).
6. Add nitrate and ammonium media to the individual Arabidopsis plant at alternate days.

3.2 Bacterial Inoculation

1. For primary bacterial culture preparation add 20 μL of frozen *P. syringae* pv. tomato DC3000 to 5 mL of KB broth containing rifampicin (50 mg/mL). Incubate on shaker at 28 °C for 24 h.
2. Take 1% of primary culture and mixed with 100 mL of KB broth containing the same antibiotic. Grow the culture till OD_{600} set to 0.6.
3. Centrifuge at 2200 × *g* for 10 min at 4 °C in Oakridge tube.
4. Discarded the supernatant and dissolve the pellet in 1 mL autoclaved 10 $mMgCl_2$ till OD_{600} set to 0.235.
5. Infiltrate the Arabidopsis leaves with suspension of *P. syringae* DC 3000 on abaxial side with the help of 1 mL syringe (*see* **Note 1**).

3.3 Fusarium oxysporum *(f. sp.* conglutinans*)* Inoculation

1. Add 39 g of Potato Dextrose Agar (PDA) to 1 L MQ.
2. Autoclave for 30 min.
3. Make the PDA slants. Cut the *Fusarium* mycelium with the help of fine blade and place it on PDA slant surface.

4. Incubate the slant at 25 °C for 7 days and allow the mycelium to grow. Store the PDA slant at 4 °C for further use.

3.4 NBT Staining Solution Preparation

1. Take the conical flask, dissolve 0.1 g NBT in 50 mM sodium phosphate buffer (pH 7.5).
2. Make up the volume with MQ to 50 mL to get a 0.2% solution (*see* **Note 2**).

3.5 NBT Staining Procedure

1. Cut the leaf from the Nitrate and Ammonium treated *Arabidopsis* plants (0 h $MgCl_2$ and 24 h *P. syringae* pv. Tomato DC 3000 Avr infiltration). Wash the leaves with distilled water.
2. Keep the leaves in 60 mm small petri dish containing NBT staining solution (0.2% solution). Wrap the 60 mm small petri dish with aluminum foil and keep it for 6 h at room temperature on shaker.
3. Drain off the NBT solution from the 60 mm small Petridish and add absolute ethanol to destain the leaves. Place the 60 mm small Petridish in a boiling water bath (~90–95 °C setting) for 10 min to remove the chlorophyll (*see* **Note 3**).
4. Visualize the leaves using Foldscope and capture the image using smartphone connected to Foldscope (Figs. 1 and 2).

3.6 Trypan Blue Staining Procedure

1. Take 10 mL lactic acid (85% w:w), 10 mL phenol (TE buffer equilibrated, pH 7.5–8.0), 10 mL glycerol, 10 mL of distilled water, and 40 mg trypan blue to prepare final concentration of 10 mg/mL trypan blue solution.

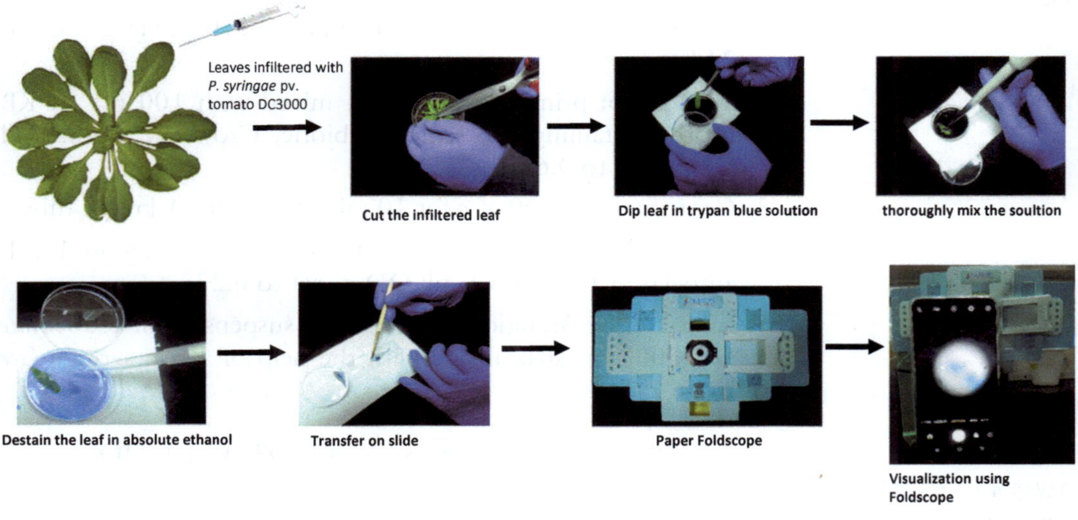

Fig. 1 Visualization of superoxide from Pseudomonas-infected leaves of Arabidopsis using Foldscope. Various steps are indicated in the figure

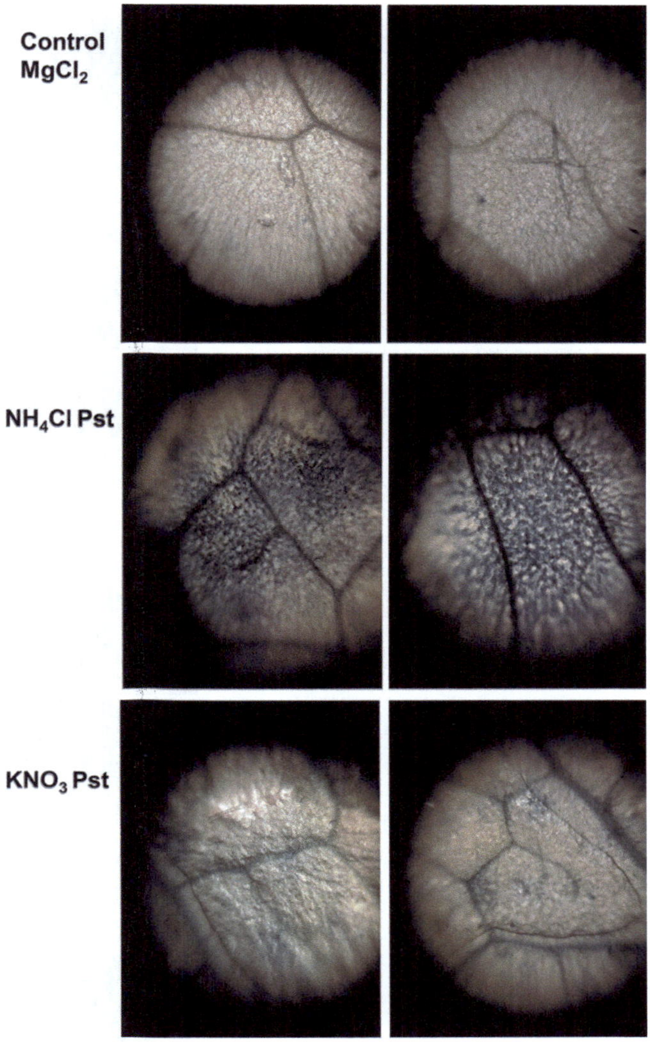

Fig. 2 Visualization of superoxide production by NBT staining. Control or NH_4^+ or NO_3^- grown leaves were infiltrated with avirulent *Pst* DC 3000 and stained with NBT solution and visualized using Foldscope and captured image using a smart phone

2. Take leaves of respective treatment and incubate them in trypan blue staining solution for 1 h.
3. Drain off the trypan blue staining solution and add absolute ethanol to destain the leaves (overnight).
4. Visualize the leaves using Foldscope (Fig. 3).

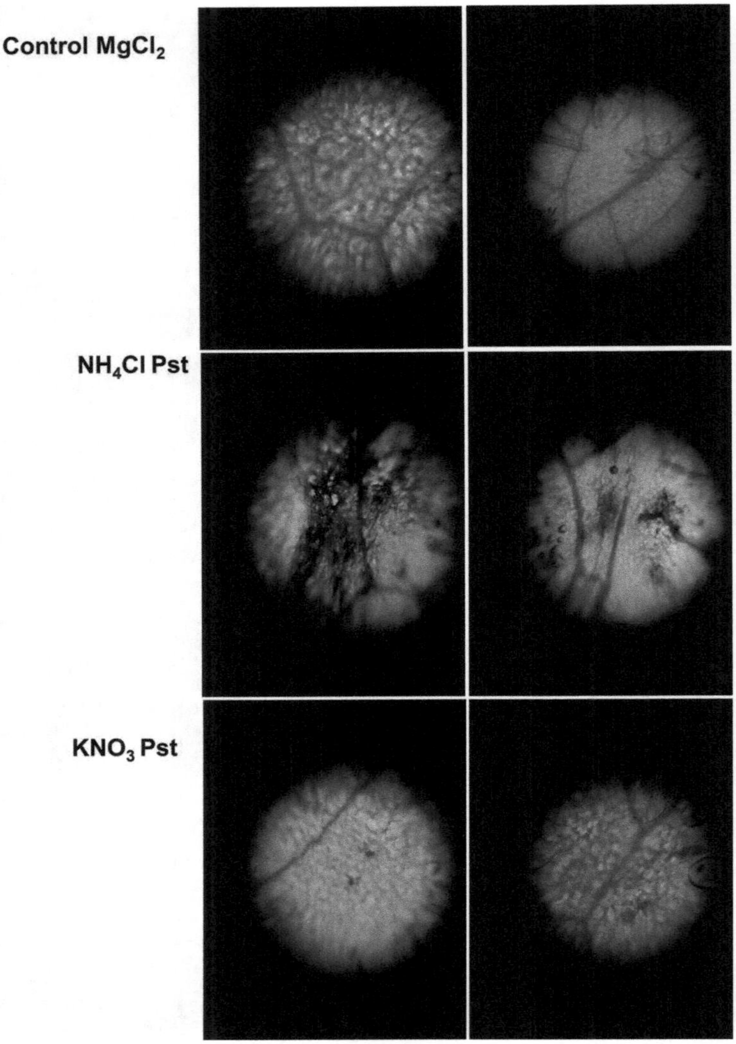

Fig. 3 Trypan blue *staining of Arabidopsis* leaves infected with avirulent Pseudomonas syringae. Leaf Images were captured by using Foldscope after 24 h of infiltration in Control or NH_4^+ or NO_3^- grown plants

4 Notes

1. Try to avoid the physical damage to the leaf during the syringe infiltration.
2. Make fresh NBT staining solution.
3. Place the 60 mm small petri dish carefully in a boiling water bath.

Acknowledgments

This work is supported by Foldscope project funded by DBT-Prakash lab initiative. PS and PKP are funded by UGC SRF and SRF.

References

1. Mittler R (2017) ROS are good. Trends Plant Sci 22:11–19
2. Schieber M, Chandel NS (2014) ROS function in redox signaling and oxidative stress. Curr Biol 24:R453–R462
3. Reczek CR, Chandel NS (2015) ROS-dependent signal transduction. Curr Biol 33:8–13
4. Delledonne M, Xia Y, Dixon RA, Lamb C (1998) Nitric oxide functions as a signal in plant disease resistance. Nature 394:585
5. Wany A, Gupta AK, Kumari MS et al (2018) Nitrate nutrition influences multiple factors in order to increase energy efficiency under hypoxia in Arabidopsis. Ann Bot 123 (4):691–705
6. De Gara L, de Pinto MC, Tommasi F (2003) The antioxidant systems vis-à-vis reactive oxygen species during plant–pathogen interaction. Plant Physiol Biochem 41:863–870
7. Tommasi F, Paciolla C, de Pinto, MC, Gara LD (2001) A comparative study of glutathione and ascorbate metabolism during germination of Pinus pinea L seeds. J Exp Bot 52:1647–1654
8. Halliwell B (1978) Lignin synthesis: the generation of hydrogen peroxide and superoxide by horseradish peroxidase and its stimulation by manganese (II) and phenols. Planta 140:81–88
9. Adam A, Farkas T, Somlyai G, Hevesi M, Kiraly Z (1989) Consequence of $O_2\cdot-$ generation during a bacterially induced hypersensitive reaction in tobacco: deterioration of membrane lipids. Physiol Mol Plant Pathol 34:13–26
10. Levine A, Tenhaken R, Dixon R, Lamb C (1994) H_2O_2 from the oxidative burst orchestrates the plant hypersensitive disease resistance response. Cell 79:583–593
11. Mittler R, Hallak Herr E, Orvar BL, Van Camp W, Willekens H, Inzé D, Ellis BE (1999) Transgenic tobacco plants with reduced capability to detoxify reactive oxygen intermediates are hyperresponsive to pathogen infection. Proc Natl Acad Sci U S A 96:14165–14170
12. Van Breusegem F, Dat JF (2006) Reactive oxygen species in plant cell death. Plant Physiol 141:384–390
13. Vanacker H, Foyer CH, Carver TL (1999) Changes in apoplastic antioxidants induced by powdery mildew attack in oat genotypes with race non-specific resistance. Planta 208:444–452
14. Mittler R, Feng X, Cohen M (1998) Post-transcriptional suppression of cytosolic ascorbate peroxidase expression during pathogen-induced programmed cell death in tobacco. Plant Cell 10:461–473
15. Vishwakarma A, Kumari A, Mur LA, Gupta KJ (2018) A discrete role for alternative oxidase under hypoxia to increase nitric oxide and drive energy production. Free Radic Biol Med 122:40–51
16. Samsatly J, Copley TR, Jabaji SH (2018) Antioxidant genes of plants and fungal pathogens are distinctly regulated during disease development in different Rhizoctonia solani pathosystems. PLoS One 13:e0192682
17. Gupta KJ, Brotman Y, Segu S, Zeier T, Zeier J, Persijn ST, Cristescu SM, Harren FJM, Bauwe H, Fernie AR (2013) The form of nitrogen nutrition affects resistance against Pseudomonas syringae pv. phaseolicola in tobacco. J Exp Bot 64:553–568
18. Modolo LV, Augusto O, Almeida IM, Magalhaes JR, Salgado I (2005) Nitrite as the major source of nitric oxide production by Arabidopsis thaliana in response to Pseudomonas syringae. FEBS Lett 579:3814–3820
19. Liszkay A, van der Zalm E, Schopfer P (2004) Production of reactive oxygen intermediates (O2−, H_2O_2, and OH) by maize roots and their role in wall loosening and elongation growth. Plant Physiol 136:3114–3123
20. Warwar N, Mor A, Fluhr R, Pandian RP, Kuppusamy P, Blank A (2011) Detection and imaging of superoxide in roots by an electron spin resonance spin-probe method. Biophys J 101:1529–1538

21. Grellet Bournonville CF, Díaz-Ricci JC (2011) Quantitative determination of superoxide in plant leaves using a modified NBT staining method. Phytochem Anal 22:268–271
22. Wohlgemuth H, Mittelstrass K, Kschieschan S, Bender J, Weigel HJ et al (2002) Activation of an oxidative burst is a general feature of sensitive plants exposed to the air pollutant ozone. Plant Cell Environ 25:717–726
23. Vleeshouwers VG, van Dooijeweert W, Govers F, Kamoun S, Colon LT (2000) The hypersensitive response is associated with host and nonhost resistance to Phytophthora infestans. Planta 210(6):853–864
24. Johansson ON, Nilsson AK, Gustavsson MB, Backhaus T, Andersson MX, Ellerström M (2015) A quick and robust method for quantification of the hypersensitive response in plants. PeerJ e1469:3
25. Planchet E, Sonoda M, Zeier J, Kaiser WM (2006) Nitric oxide (NO) as an intermediate in the cryptogein-induced hypersensitive response–a critical re-evaluation. Plant Cell Environ 29:59–69
26. Keogh RC, Deverall BJ, McLeod S (1980) Comparison of histological and physiological responses to Phakopsora pachyrhizi in resistant and susceptible soybean. Trans Br Mycol Soc 74(2):329–333
27. Bouarab K, Melton R, Peart J, Baulcombe D, Osbourn A (2002) A saponin-detoxifying enzyme mediates suppression of plant defences. Nature 418(6900):889
28. Zhou L, Cheung MY, Li MW, Fu Y, Sun Z, Sun SM, Lam HM (2010) Rice hypersensitive induced reaction protein 1 (OsHIR1) associates with plasma membrane and triggers hypersensitive cell death. BMC Plant Biol 10(1):290
29. Pogany M, von Rad U, Grün S, Dongo A, Pintye A, Simoneau P et al (2009) Dual roles of reactive oxygen species and NADPH oxidase RBOHD in an Arabidopsis-Alternaria pathosystem. Plant Physiol 151(3):1459–1475

Chapter 10

Methods for Estimation of Nitrogen Components in Plants and Microorganisms

Pratiksha Singh, Rajesh Kumar Singh, Qi-Qi Song, Hai-Bi Li, Li-Tao Yang, and Yang-Rui Li

Abstract

Nitrogen (N_2) is the most necessary element in the atmosphere, it is an energetic micronutrient for plant growth and development after water, besides its key role in chlorophyll production, which is crucial for photosynthesis process. Biological nitrogen fixation is measured to be the most potent method to deliver a fixed way of nitrogen to the plants. Plant depends on free-living and symbiotic microbes present in the soil for nitrogen because it cannot be absorbed by the plant itself directly from the atmosphere. Many techniques were reported in the laboratory for nitrogen estimation till now, but Kjeldahl digestion and acetylene reduction assay (ARA) techniques became the most popular. In this chapter, we focus on the most common and popular methods used to determine plant N_2; awareness obtained through the wide application of these methods should offer the source for the N_2 fixation rate in agriculture system.

Key words Acetylene reduction assay, Combustion method, Kjeldahl, Microbial bioassay, $^{15}N_2$

1 Introduction

Nitrogen (N_2) is an essential nutrient that is required to all living being on Earth, and it is also necessary for the production of nucleic acids, amino acids, chlorophyll, enzymes, and proteins including DNA for growth. In 1886, German scientists Hellriegel and Wilfarth discovered N_2 fixation and stated that legumes bearing root nodules can use gaseous (molecular) nitrogen. Then, in 1988, Beijerinck isolated a bacterial strain from root nodules which is known as *Rhizobium leguminosarum* [1]. Biological nitrogen fixation is accountable for N_2 fixation of complicated nitrogenase enzyme, which comprises three subunits and is controlled by a complex arrangement with several genes [2]: nitrogenase I, nitrogenase II, and nitrogenase III. Nitrogenase I is dependent on iron and molybdenum encoded by *nif* gene, nitrogenase II is vanadium encoded by *vnf* gene, and nitrogenase III is iron encoded by *anf* gene [3, 4]. Out of these *nif* genes encodes conserved subunits in

Kapuganti Jagadis Gupta (ed.), *Nitrogen Metabolism in Plants: Methods and Protocols*, Methods in Molecular Biology, vol. 2057, https://doi.org/10.1007/978-1-4939-9790-9_10,

free-living and associated microorganisms [4–6]. The *nif*genes include *nif*D, *nif*H, and *nif*K, which totally encodes proteins of the nitrogenase enzyme complex [2]. There are several methods evolved for the estimation of nitrogen from soil and plant. In laboratory, several methods evolved for nitrogen estimation by living organisms. According to Hardy et al. [7], some of them are Kjeldahl analysis and ^{15}N-enrichment by mass spectrometry [8], $^{13}N_2$-incorporation by radioactive counting [9, 10], N_2 fixation by nitrogenase in cell-free extracts by $^{15}N_2$ enrichment [11], Conway microdiffusion technique with titrimetric [12] colorimetric analysis of NH_3 [13], and N_2–H_2 uptake [14]. In this chapter, we discuss some popular conventional methods.

1.1 Kjeldahl Analysis [15]

This method is known as Kjeldahl digestion, and it has been widely used for nitrogen determination on food, beverages, meat, feed, grain, manure, wastewater, soils, and plant tissue. It is considered a method of reference for biological sample nitrogen estimation [16] (Fig. 1a).

1.2 $^{15}N_2$ Abundance Determination (Mass Spectrometry)

$^{15}N_2$ abundance determination and $^{15}N_2$ isotope dilution is a recent, rapid, and highly sensitive method for determination of N_2 in plants and other biological samples, as well as applicable for low and small sample volumes. This method is based on the theory that the concentration of $^{15}N_2$ in the atmosphere is different from the plants and thus the analysis of $^{15}N_2$ of the nitrogen fixing and nonfixing plant is measured as the amount of nitrogen fixed [1] (Fig. 1b).

1.3 Dumas Combustion Method

This method of total nitrogen determination in organic and inorganic samples was proposed by Jean-Baptiste Dumas in 1831 [17].

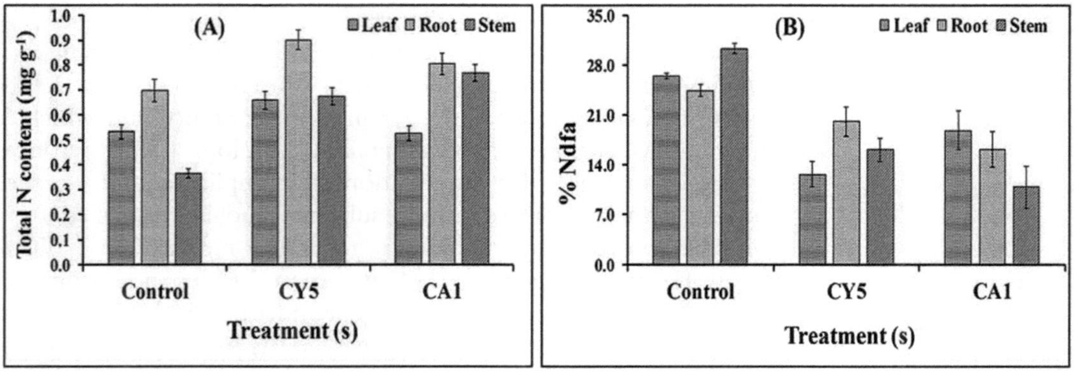

Fig. 1 (**a**, **b**) Effect of inoculation of strains *Bacillus megaterium* (CY5) and *Bacillus mycoides* (CA1) on N and percent ^{15}N parameters for dry biomass of sugarcane varieties GT11. The columns represent the mean of the data for each treatment and bars represent the standard error

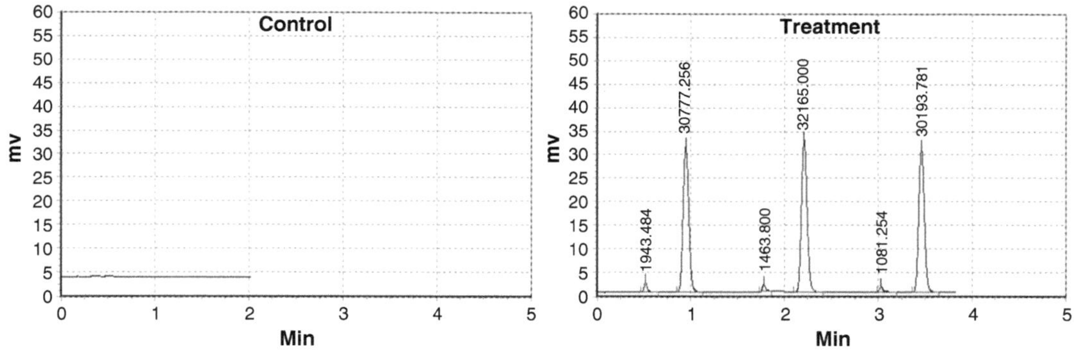

Fig. 2 Acetylene reduction assay for control and treatment with potent strain showing different peaks in three replicates

1.4 Acetylene Reduction Assay

Quantification of N_2 fixation by nitrogen-fixers is essential for knowing their involvement in total nitrogen budget in a particular ecosystem [18]. Dilworth [19] provided a useful assay for quantification of N_2 fixation rate which relies on the preferential reduction of acetylene (C_2H_2) to ethylene (C_2H_4) by nitrogenase, instead of reducing N_2 to NH_3 [20]. Here, we discuss the method of acetylene reduction assay which was modified by Castle [21] (Fig. 2).

1.5 Microbial Bioassay

Das and De [22] proposed a new method "microbial bioassay" for determination of nitrogen fixation rate and in this technique nitrogen fixers are cultivated in a specific medium, after required time of interval the amount of nitrogen fixed by them is calculated.

2 Materials

2.1 Kjeldahl Analysis

1. Azotometer (K-05 type automatic azotometer).
2. Digestion apparatus (X20A aluminum module automatic digestion device).
3. The catalyst consists potassium sulfate, copper sulfate, and selenium powder in a ratio of 100:10:1.
4. Indicator: 0.5 g of bromocresol green and 0.1 g of methyl red were dissolved in 100 mL of absolute ethanol to prepare an indicator liquid.
5. Boric acid absorption liquid: 5 mL of indicator stock solution was added in 500 mL of 2% boric acid solution.
6. Concentrated sulfuric acid (98%).
7. Sodium hydroxide solution (40%): 400 g of NaOH dissolved in 1000 mL of distilled water (*see* **Note 1**).

2.2 $^{15}N_2$ Abundance Determination (Mass Spectrometry)

1. Mass spectrometer Thermo-Fisher Delta V Advantage IRMS, measured by DI measurement.
2. Lithium hypobromide solution: 90 g of lithium hydroxide is weighed and dissolved in 1000 mL of distilled water. The mixture is placed in ice water, and 30 mL of liquid bromine (density 3.12 g mL^1) is added slowly, and the mixture is shaken to complete the reaction, refrigerate after overnight (*see* **Note 2**).

2.3 Dumas Combustion Method

1. Thermoelectric cooler.
2. Reduction heater.
3. Thermal conductivity detector (*see* **Note 3**).

2.4 Acetylene Reduction Assay

1. A 24-in. section of Tygon tubing with a needle fitting.
2. 1000 mL volumetric flask with armhole and appropriately sized rubber stopper.
3. The tedlar gas collection bag.
4. 5 cm PVC coring cylinders with yellow caps.
5. 500 mL canning jar with lid bottom drilled hole.
6. One hole rubber stopper fitted with glass tubing.
7. Septa Cyl Half-hole 1/4 in. 100 pk.
8. 1000 μL pipette.
9. 60 mL gastight syringes with two-way stopcocks.
10. 22 gauge syringe needles.
11. Labco Exetainer sample tubes (pre-evacuated).
12. Calcium Carbide (CaC_2) (1 g CaC_2 = 130 mL of C_2H_2 gas; or ~0.3 g per sample).
13. Nanopure DI water (*see* **Note 4**).

2.5 Microbial Bioassay

1. Phosphate buffer isotonic solution.
2. Jensen's medium (Composition: sucrose 20 g L^{-1}, dipotassium phosphate 1 g L^{-1}, magnesium sulfate 0.5 g L^{-1}, sodium chloride 0.5 g L^{-1}, ferrous sulfate 0.1 g L^{-1}, sodium molybdate 0.005 g L^{-1}, calcium carbonate 2 g L^{-1}, agar 15 g L^{-1} (*see* **Note 5**).

3 Methods

3.1 Kjeldahl Analysis

1. This method can be completed in three steps:

   Wet digestion → Distillation → Ammoniuam estimation

2. The plants samples (0.1–0.3 g; depending on different varieties, different parts of plants, and different periods) and the

soil samples (1–5 g) are accurately weighed. Keep it inside the digestive tube.

3. Add 1 g of catalyst in each sample (potassium sulfate–copper sulfate–selenium powder; 100:10:1).
4. Add 5 mL of concentrated sulfuric acid to the plant sample, while 8–10 mL of concentrated sulfuric acid is added to the soil sample.
5. Place the digestive tube on a digestion apparatus and set the temperature to 410 °C. After the complete digestion of plant samples, continue digestion for 2 h; however, after digestion of the soil samples, continue digestion for 4 h.
6. Add 20–30 mL of 40% sodium hydroxide solution in the digestive tube; carry out distillation and absorb the distillate in 10 mL of 2% boric acid solution, and the indicator is methyl red-bromocresol green. The distillation time is 6 min.
7. Titrate with calibrated sulfuric acid and record the titration volume.
8. Add one drop of 1N sulfuric acid in the triangular flask, then concentrate on an electric heating plate and concentrate to about 1 mgN mL^{-1} to measure the abundance of $^{15}N_2$.

   Calculation:

   $$N\% = 1.401 \times N \times (V - V_0)/W$$

   Blank titration volume: V_0 (mL); sample titration volume: V (mL); standard acid equivalent concentration N; sample weight W (g).

3.2 $^{15}N_2$ Abundance Determination (Mass Spectrometry)

1. Add 1 mL (about 1 mg NmL^{-1}) of the concentrated sample after Kjeldahl method to the side of the *Υ*-shaped bottle, then add 1 mL of lithium hypobromide solution to the other side of the *Υ*-shaped bottle, freeze with liquid nitrogen, and evacuate to 2×10^{-4} mBar, thaw, mix the reaction liquid on both sides of the *Υ*-shaped bottle to convert NH_4^+ into nitrogen. The chemical reaction is as follows:

   $$2NH_4^+ + 3BrO^- + 2OH^- \rightarrow N_2 + 3Br^- + 5H_2O$$

2. The generated N_2 enters the ion source ionization under high vacuum conditions (1×10^{-7} mBar), converts N_2 to $[N_2]^+$, and records the $[^{28}N_2]^+$, $[^{29}N_2]^+$, $[^{30}N_2]^+$ ion peak intensity.
3. Sample ^{15}N Atom% ($\leq$10%) is calculated as follows:

   $$R = [^{28}N_2]^+ / [^{29}N_2]^+$$

   $$^{15}N \text{ Atom\%} = \frac{1}{2 \times R + 1} \times 100\%$$

4. Sample $^{15}N\%$ (>10%) is calculated as follows:

$$^{15}\text{N Atom\%} = \frac{[^{29}N_2{}^+] + 2 \times [^{30}N_2]^+}{2 \times \left([^{28}N_2]^+ + [^{29}N_2]^+ + [^{30}N_2]^+\right)} \times 100\%$$

3.3 Dumas Combustion Method

1. Weigh 200–300 mg sample into a tin capsule.
2. Heat a tin capsule containing a sample at 800–1000 °C temperature.
3. After combustion, some gases such as carbon dioxide, oxygen, nitrogen oxides, gaseous nitrogen, and water vapor, are generated as a by-product. Can remove water vapor by different water removal methods (e.g., perchlorate trap) [23], by passing gas products through a thermoelectric cooler [24] or an equivalent.
4. Reduce nitrogen oxides into gaseous nitrogen N_2. This can be done by passing gasses through pure copper fillings in a reduction heater.
5. Separate remnant gases, such as carbon dioxide and gaseous nitrogen, and trap CO_2 to measure only the gas nitrogen concentration.
6. In the last step, measure the nitrogen concentration through a thermal conductivity detector by using differences in thermal conductivity of the gases.

3.4 Acetylene Reduction Assay

1. Acetylene gas is prepared by mixing a suitable amount of calcium carbide rocks and ½ cup of water in a flask. Rapidly cover flask opening with a rubber stopper let flask vent for several seconds before inserting syringe needle into the collection bag. Make sure that needle does not puncture the gas bag. Allow bag to fill with acetylene. Take out the needle from bag and flask when completed. Place flask in the hood or allow flask to vent out the window until the reaction is complete.

$$CaC_2 + H_2O \rightarrow C_2H_4$$

2. Sample "wet-up" is done by adding a suitable volume (Vwater) of nanopure DI water to 5 cm sample core by a pipette. Place soil core into incubation chamber for 4 h. Chambers conditions: On all lights (88), set temperature and humidity 10%.
3. Create 10% acetylene atmosphere by placing soil cores into incubation jars and stops the jar. Remove 38 mL of air from the jar (for a 500 mL Mason jar has a full 5 cm core volume) with a gastight syringe. Then inject 38 mL acetylene into the jar and let the acetylene equilibrate with the atmosphere by venting the jar with a needle so the jar is not overpressurized.

4. Push the air in the jar smoothly with a syringe to mix the acetylene in the jar and remove 24 mL from the sample jars through a gastight syringe and insert into a pre-evacuated Labco Exetainer. Exetainer vials should be overpressurized. Record the time (t_0) during sampling.
5. Return sample jars to the incubator at the same settings as "wet-up" incubation for the time period of interest (generally 3 h). Again, remove 24 mL from jars and inject into Exetainer. Record the time (t_f).

Calculations:

1. Calculate $\Delta t = t_f - t_0$.
2. Assuming a particle density of 2.6 g mL^{-1}, calculate the volume of solids: $V_{solid} = W_{OD}/2.6$.
3. Calculate the headspace volume: $V_{headspace} = V_{total} - V_{water} - V_{solid}$.
4. Calculate the ethylene concentrations for the t_0 and t_f measurements from the calibration:
   µmol C_2H_4 $mL^{-1} = a + bx$ (x is the peak area from the gas chromatograph, a and b are derived from the calibration curve).
5. Calculate the Δmol of ethylene in the jar at t_0 and t_f: $E_{total} = (\Delta mol\ C_2H_4\ mL^{-1}) \times V_{headspace}$
   (For simplicity, we are ignoring any C_2H_4 dissolved in pore water).
6. Calculate the rate of acetylene reduction to ethylene: Rate $= [(E_{total})t_f - (E_{total})t_0)]/(\Delta t\ x W_{OD})$.

3.5 Microbial Bioassay

1. Homogenize soil sample with the sterilized phosphate buffer isotonic solution.
2. Filter soil suspension through a membrane filter (pore size, 0.8 mm).
3. The filtrate contains only bacterial cells after filtration. Make serial dilution up to 10^{-4} using sterilized phosphate buffer isotonic solution and inoculate 1 mL of this diluted bacterial suspension into 30 different petri plates containing 15 mL of nitrogen-free medium (Jensen's medium) [25].
4. Incubate plates at 30 °C under an N_2 atmosphere of 5 ppmv of concentration in an isolated incubation chamber for 14 days. Obtain pure colonies by repeating subculture through streaking on Jensen's medium.
5. Subculture morphologically different colonies for ten generations to confirm the plates contain only free-living N_2-fixing bacteria. Isolate pure culture obtained from the tenth generation from the medium with phosphate buffer saline solution.

6. Make serial dilution up to 10^{-4} from this bactertial suspension by using sterilized phosphate buffer isotonic solution and inoculate 1 mL of this solution on petri plates containing 15 mL of Jensen's medium then incubate under atmospheric N_2. Repeat this two times. Monitor petri plates at an interval of 12 h to count the increase in the number of colony-forming units.

$$\text{Total bacteria cell} = \text{cfu} \times \text{number bacteria cell present in that colony}$$

7. Monitor each colony by hemocytometer for quantification of free-living N_2-fixing bacteria.
8. Calculate total generated bacterial cell by multiplication of colony number and the bacterial cell found in the respective colony.
9. From the replicate, analyze petri plates microbial sample for quantification of nitrogen present in their cell by the high-temperature catalytic oxidation (HTCO) method [26].
10. Get the N_2 fixation rate by the amount of N_2 fixed by the free-living N_2-fixing bacteria divided by the incubation period.

    Calculation:

$$\text{Rate of } N_2 \text{ fixation} = \frac{(\text{amount of N in bacterial cell})_{\text{after 2 weeks}} - (\text{amount of N in bacterial cell})_{\text{initial}}}{(\text{Incubation period}) \times \text{mass of dry weight of soil}}$$

4 Notes

1. Kjeldahl analysis.

   All reagents prepared in autoclave bottle with HPLC grade water.

   (a) 2% boric acid absorbing solution is containing an indicator.

   (b) During acidification, the acid addition should not be too much, otherwise, bromine gas will begenerated during mass spectrometry, and the ion source will be contaminated.

$$Br^- + OBr^- + 2H^+ = Br_2 + H_2O$$

2. $^{15}N_2$ abundance determination (mass spectrometry).

   (a) High sensitive expensive and accurate method requires specialized equipment and skills; therefore each step should be done carefully.

3. Dumas combustion method.

   (a) Weighing of a sample should be done carefully due to the requirement of a small sample weight in Dumas combustion method.

(b) Pure oxygen is added to accelerate combustion.

(c) The final product is N_2 instead of ammonia.

4. Acetylene reduction assay.

(a) Take care to evenly disperse water over the entire surface of the sample core without disturbing the surface.

(b) Samples can be stored in Exetainer vials for a long time. But it is suggested that you run the samples as soon as possible.

(c) Injecting ethylene standard into vial will also help to identify whether or not vials are leaking over time.

5. Microbial bioassay.

(a) The filtration process is done for the separation of bacterial cells from other microbes because soil bacterial cells may achieve a maximum of 0.5 mm in diameter [27].

References

1. Shin W, Islam R, Benson A, Joe MM, Kim K, Gopal S, Samaddar S, Banerjee S, Sa T (2016) Role of Diazotrophic Bacteria in biological nitrogen fixation and plant growth improvement. Korean J Soil Sci Fert 49:17–29
2. Figueiredo MVB, Mergulhao ACES, Sobral JK, Junior MAL, Araujo A (2013) Biological nitrogen fixation: importance, associated diversity, and estimates. In: Arora NK (ed.) Plant microbe symbiosis: fundamentals and advances. https://doi.org/10.1007/978-81-322-1287-4_10, © Springer India
3. Franche C, Lindstro MK, Elmerich C (2009) Nitrogen-fixing bacteria associated with leguminous and non-leguminous plants. Plant Soil 321:35–59
4. Canfield D, Glazer AN, Falkowski PD (2010) The evolution and future of earth's nitrogen cycle. Science 330:192–196
5. Zehr JP, Jenkins BD, Short SM, Steward GF (2003) Nitrogenase gene diversity and microbial community structure: a cross-system comparison. Environ Microbiol 5:539–554
6. Falkowski PG, Fenchel T, Delong EF (2008) The microbial engines that drive Earth's biogeochemical cycles. Science 320:1034–1039
7. Hardy RWF, Knight E Jr (1968) The biochemistry and postulated mechanisms of N_2 fixation. In: Reinhold L (ed) Progress in phytochemistry. John Wiley and Sons Ltd, Sussex, England, pp 387–469
8. Burris RH, Wilson PW (1957) Methods for measurement of nitrogen fixation. In: Colowick SP, Kaplan NO (eds) Methods in enzymology, vol 4. Academic Press, New York, pp 355–367
9. Campbell NE, Dular R, Lees H (1967) The production of $^{13}N_2$ by 50 Mev protons for use in biological nitrogen fixation. Can J Microbiol 13:587–599
10. Nicholas DJD, Silvester DJ, Fowle JF (1961) Use of radioactive nitrogen in studying nitrogen fixation in bacterial cells and their extracts. Nature 189:634–636
11. Carnahan JE, Mortenson LE, Mower HF, Castle JE (1960) Nitrogen fixation in cell-free extracts of *Clostridium pasteurianum*. Biochim Biophys Acta 44:520–535
12. Mortenson LE (1961) A simple method for measuring nitrogen fixation by cell-free enzyme preparations of *Clostridium pasteurianum*. Anal Biochem 2:216–220
13. Dilworth MJ, Subramanian D, Munson TO, Burris RH (1965) The adenosine triphosphate requirement for nitrogen fixation in cell-free extracts of *Clostridium pastcurianum*. Biochim Biophys Acta 99:486–503
14. Mortenson LE (1964) Ferredoxin and ATP, requirements for nitrogen fixation in cell-free extracts of *Clostridium pasteurianum*. Proc Natl Acad Sci U S A 52:272–279
15. Kjeldahl J (1883) Neue methode zur bestimmung des stickstoffs in organischen körpern. Fresenius J Anal Chem 22:366–382
16. Labconco CA (1998) Guide to Kjeldahl nitrogen determination methods and apparatus. Labconco cCorporation, Houston

17. Dumas JBA (1831) Procedes de l'analyse organic. Ann Chim Phys 247:198–213
18. Stem WR (1993) Nitrogen fixation and transfer in intercrop systems. Field Crop Res 34:335–356
19. Dilworth MJ (1966) Acetylene reduction by nitrogen-fixing preparations from *Clostridium pasteurianum*. Biochem Biophys Acta 127:285–294
20. Crews TE, Kurina LM, Vitousek PM (2001) Organic matter and nitrogen accumulation and nitrogen fixation during early ecosystem development in Hawaii. Biogeochemistry 52:259–279
21. Castle S (2010) Acetylene reduction assay (ARA): measuring Nitrogenase activity. Aridlands ecology lab protocol modified: 2010.02.16
22. Das S, De TK (2018) Microbial assay of N_2 fixation rate, a simple alternate for acetylene reduction assay. MethodsX 5:909–914
23. Unkovich M, Herridge D, Peoples M, Cadisch G, Boddey B, Giller K, Alves B, Chalk P (2008) Measuring plant-associated nitrogen fixation in agricultural systems. Australian Centre for International Agricultural Research (ACIAR), Canberra, ACT, Australia, pp 45–62
24. Lee D, Nguyen V, Littlefield S (1996) Comparison of methods for determination of nitrogen levels in soil, plant and body tissues, and water. Commun Soil Sci Plant Anal 27:83–793
25. Kayasth M, Gera R, Dudeja SS, Sharma PK, Kumar V (2014) Studies on salinization in Haryana soils on free-living nitrogen-fixing bacterial populations and their activity. J Basic Microbiol 54:170–179
26. Lee KK, Watanabe I (1977) Problems of the acetylene reduction technique applied to water- saturated paddy soils. Appl Environ Microbiol 34:654–660
27. Portillo MC, Leff JW, Lauber CL, Fierer N (2013) Cell size distributions of soil bacterial and archaeal taxa. Appl Environ Microbiol 79:7610–7617

Chapter 11

A Precise Method for Analyzing Nitrogen Use in Foxtail Millet

Tirthankar Bandyopadhyay and Manoj Prasad

Abstract

Optimization of biological nitrogen (N) use is instrumental in ensuring higher crop yields and preventing environmental degradation due to excessive N fertilizer application. Furthermore, understanding how genetic differences differentially influence N remobilization into seeds under contrasting nitrogen nutrition regimes is crucial to our understanding of nitrogen use efficiency (NUE) in crops in addition to enabling a deeper mechanistic understanding of the dynamics of nitrogen metabolism in plants. In this chapter, a method is proposed to precisely measure and analyze nitrogen use efficiency (NUE) in a pot-based system under different nitrogen nutrition regimes in foxtail millet (*Setaria italica L.*), a climate change-resilient C_4 model crop with great promise for food security and nutrition in the twenty-first century.

Key words Foxtail millet, C_4 crops, Nitrogen use efficiency, Dumas method, Yield per plant

1 Introduction

The aim to reduce N fertilizer input and breed plants with enhanced NUE is one of the major goals of plant nutrition research globally [1, 2]. A precise analysis of NUE traits especially under low and sufficient N conditions and to account for the allelic variation in N use in crops requires a growth system that enables accurate measurements of major parameters determining the same. We propose a pot-based full-cycle growth system that is ideal for analyzing NUE, NRE at the metabolic, genomic and physiological levels, thereby facilitating the establishment of a robust system for understanding nitrogen metabolism in this important C4 model crop. The method has certain specific advantages: higher control of N provisioning; calculation of N uptake by the plant; a relatively N inert growth system; no N leaching; portability of the experimental set up for high-throughput phenotyping. The present method is based on successful N stress phenotyping optimization at five different N growth conditions for *Setaria italica* accession IC-480117 (IC-41). The method aims to estimate major NUE

Kapuganti Jagadis Gupta (ed.), *Nitrogen Metabolism in Plants: Methods and Protocols*, Methods in Molecular Biology, vol. 2057, https://doi.org/10.1007/978-1-4939-9790-9_11,

related traits in the crop, namely, NUE [3] as well as to analyze seed N content (%) (Dumas method).

2 Materials

1. Growth Substrate: Soilrite mix–vermiculite (2:1 w/w). Mix thoroughly, keep dry.
2. Plastic Pots: 19.5 cm high × 20 cm diameter. Fill 2.5 kg of dry substrate mix in each pot.
3. Seeds for the experiment.
4. Nutrient solution: Modified Hoagland solution with five different levels of Ca $(NO_3)_2{\cdot}4H_2O$ (Table 1).

3 Methods

1. Growth system: Foxtail millet, being an arid annual crop needs drier and warmer growing conditions. Ideally, the seeds can be sowed in April/May and seeds harvested in July/August each year. Therefore, the greenhouse conditions were maintained at 30–35 °C/20–25% RH/10:14 photoperiod with 350 μmol/m^2/s illumination.
2. Substrate for plant growth: Plastic pots (19.5 cm high × 20 cm diameter) were filled with 2.5 kg of the Soilrite mix–vermiculite (2:1 w/w; nutrient and nitrogen free) and soaked in 1500 ml of demineralized (DM) water for 1 day.
3. Seed treatment and sowing: Seeds were soaked in 2 g/l of Mancozeb 75% WP broad-spectrum fungicide for 20 min, air dried and sown (between 09:00 and 12.00 h) in pots (5 seeds/pot) and left for germination for next 6 days. Accessions were accessed for germination on the 7th day of sowing (DOS), watered (300 ml RO) and observed for vegetative growth for the next 7 days. Four plants per pot with uniform and similar vegetative growth were retained for subsequent study.
4. N dosage: Plants at 14 DOS or 7 days after germination(DAG) were initiated into fertigation with 500 ml of modified Hoagland solution [4] prepared in demineralized water with three varying concentrations of $Ca(NO_3)2$ at 2000 μM (full nutrition, control), 500 μM (25% N, low N) and 200 μM (10%, very low N) (Table 1). Pots were fertigated once every week periodically between 16.00 and 17.30 h till seed maturity (17 weeks).
5. Upon maturity (Figs. 1 and 2), shoots (including the panicle) and roots (after carefully removing the growth substrate by gently rinsing them in RO water followed by tissue drying)

Table 1
Modified Hoagland nutrient solution

Component	Mol. mass	Strengthof 1× (mM)	Stock strength	Amount for respective stock (g) in 1 l
1. KH_2PO_4	136.09	1	1000×	136.08
2. KCl[a]	74.55	4.9	400×	372.7
3. $MgSO_4$	120.7	1.99	1000×	240.32
4. $MnCl_2$	197.91	0.009	1000×	1.81
5. H_3BO_3	61.83	0.0462	1000×	2.86
6. Na_2MoO_4	241.95	0.0001	1000×	0.025
7. $CuCl_2$	170.48	0.0026	1000×	0.045
8. $ZnCl_2$	136.28	0.00807	1000×	0.11
9. EDTA ferric monosodium salt	367.05	0.089	1000×	33.0
10. Ca $(NO_3)_2{\cdot}4H_2O$	236.5 236.5 236.5 236.5 236.5	2 (N100) 1.0 (N50) 0.5 (N25) 0.2 (N10) 0.0 (N0)	1000× – – –	473 Diluted from stock for 2 mM) Diluted from stock for 2 mM) Diluted from stock for 2 mM –
11. $CaCl_2{\cdot}2H_2O$[a]	147.02 147.02 147.02 147.02 147.02	2.07(N100) 3.0 (N50) 3.5 (N25) 3.9 (N10) 0.0 (N0)	– – – 500× –	Diluted from stock for 3.9 mM Diluted from stock for 3.9 mM Diluted from stock for 3.9 mM 573.37 —

[a]1000× KCl and $CaCl{\cdot}2H_2O$ stock solution cause precipitation, therefore 400× and 500× of the same were prepared, respectively (*see* **Note 1**)

Fig. 1 Mature foxtail millet plants grown at five N dosages. Full-cycle growth (17 weeks) experiment was carried out under controlled conditions to optimize nutrient dosage and growth condition. N0, N10, N25, N50, and N100 represent 0, 10, 25, 50, and 100% of the full dosage of $Ca(NO_3)_2{\cdot}4H_2O$(2 mM) as part of modified Hoagland nutrient solution

Fig. 2 Panicle morphology and size after maturity grown at maturity five when grown under five different N levels as indicated

were sun-dried individually for 2 days and weighed as shoot dry weight (SDW) and root dry weight (RDW) (Fig. 3).

6. The yield per plant (YPP) was measured as the average total dry weight (g) of seeds per plant per treatment from four biological replications (Fig. 3).
7. Hundred seeds per plant per treatment were counted by using the programmable seed counter (Indosaw, Model 6709) and subsequently weighed.
8. 500 mg of seeds from each sample was manually crushed into fine dry powder using mortar and pestle, labeled and processed for N and C content analysis (as percentage of dry weight) as per the standard protocol for CHN(S) elemental analyzer (Thermo Finnigan, FLASH EA 1112 series, Italy) using Dumas method (Fig. 4). The analysis was carried out at three N conditions (N10, N25, and N100) to obtain an idea of variation of N content across five N conditions (N0 and N50 were excluded from the study).
9. The nitrogen use efficiency (NUE) for each treatment was calculated as

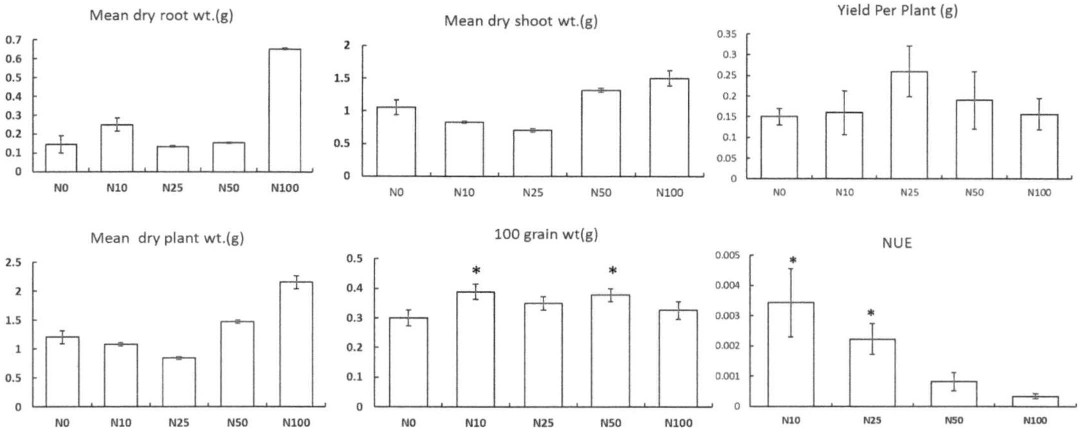

Fig. 3 Mean dry weight of root, shoot, YPP, total plant weight, 100-grain weight and NUE at maturity as indicated. N0, N10, N25, N50, and N100 represent 0, 10, 25, 50, and 100% of the full dosage of $CaNO_3{\cdot}4H_2O$ (2 mM) as part of the Hoagland nutrient solution. *Error bars* represent standard deviation *while Bars* with *astrix* (∗) represent significance at $\alpha \leq 0.05$ against control (N100) calculated using students *t*-test. Values in *Y*-axis for all parameters except NUE (values) represent weight in grams. All data except NUE are a mean of four individual replications. Data for NUE are a mean of three replications

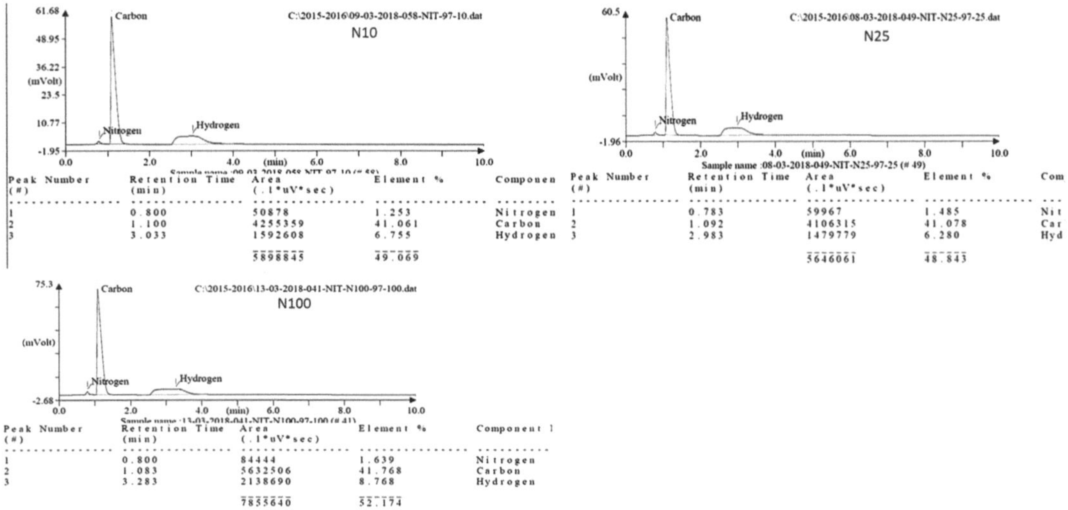

Fig. 4 CHN analysis output indicating C, N and H content (in percentages of unit sample weight) from N10, N25, and N100 using Dumas method. K factor-based calibration method was used for the analysis with a constant protein factor of 6.25

Total seed weight (g) per plant (YPP)/total N supplied [5] as well as (N% seeds × DW seeds)/N supplied (Griffith Lab).

The amount of N supplied per dose is calculated as follows:

(a) Amount of Ca $(NO_3)_2{\cdot}4H_2O$ in 2 mM Solution of volume 1 l = (Mol. Wt/1000) × 2.

(b) Mole Fraction of N in Ca $(NO_3)_2{\cdot}4H_2O$ = Atomic weight of N in $Ca(NO_3)_2{\cdot}4H_2O$/Molecular weight of Ca $(NO_3)_2{\cdot}4H_2O$.

(c) Multiply a × b to get the amount of N in 1 l of Ca $(NO_3)_2{\cdot}4H_2O$ in 2 mM.

(d) Divide the value obtained in (c) above by 2 (500 ml applied per pot) to get the amount of N applied per dose of N100 (2 mM Ca $(NO_3)_2{\cdot}4H_2O$ in 2 mM in Hoagland solution). Calculate for other doses (N50, N25, N10, and N0) accordingly.

4 Notes

1. 1000× stock solution for KCl and $CaCl{\cdot}2H_2O$ collects precipitate, so prepare 400× and 500× stocks, respectively (Table 1). Stocks and working solution to be prepared in deionized water and pH of the working solution (1×) adjusted to 5.8 ± 0.3 using 5 N NaOH or HCl and stored at room temperature. EDTA ferric monosodium salt to be stored in an amber-colored bottle. Modified from [4], refer to component list in Table 1.

Acknowledgments

The authors' work in this area is supported by DBT-BBSRC collaborative project (File no. BT/IN/UK-VNC/42/RG/2015-16). The authors also thank Mr. Anand Dangi, Technical Staff, NIPGR, New Delhi for his assistance.

References

1. Hirel B, Le Gouis J, Ney B, Gallais A (2007) The challenge of improving nitrogen use efficiency in crop plants: towards a more central role for genetic variability and quantitative genetics within integrated approaches. J Exp Bot 58:2369–2387
2. Han M, Okamoto M, Beatty PH, Rothstein SJ, Good AG (2015) The genetics of nitrogen use efficiency in crop plants. Annu Rev Genet 49:269–289
3. Moll RH, Kamprath EJ, Jackson WA (1982) Analysis and interpretation of factors which contribute to efficiency to nitrogen utilization. Agron J 74:562–564
4. Hoagland DR, Arnon DI (1950) The water-culture method for growing plants without soil. California Agr Exp Stat Cir 347:1–32
5. Dumas JBA (1950) Procedes de l'analyse organic. Ann Chim Phys 1831(247):198–213

Chapter 12

Methods for Isolation and Characterization of Nitrogen-Fixing Legume-Nodulating Bacteria

Nisha Tak, Garima Bissa, and Hukam S. Gehlot

Abstract

Symbiotic nitrogen fixation (SNF) is a characteristic feature of nodulating legumes. The wild legumes are comparatively less explored for their SNF ability; hence, it is essential to study nodulation and identify the microsymbiont diversity associated with them. This chapter aims to describe the methodology for nodule hunting; trapping, isolation, and characterization of root nodule bacteria (RNB) at phenotypic, genotypic, and symbiotic levels. The documentation of nodulating native legume species and the rhizobial diversity associated with them in various parts of world has gained attention as this symbiotic association provides fixed nitrogen, improves productivity of plants in an ecofriendly manner. Before field-based applications the symbiotic bacteria need to be assessed for their N fixing ability as well as characterized at molecular level. The phylogeny based on symbiosis-essential genes supplemented with the host-range studies helps in better understanding of the symbiotaxonomy of rhizobia. More efficient symbiotic couples need to be screened by cross-nodulation studies for their application in agricultural practices.

Key words Legume root-nodules, Rhizobia, DNA fingerprinting, Housekeeping and symbiotic genes, Phylogeny, Host-range

1 Introduction

The Leguminosae (Fabaceae) is the third largest family of flowering plants with about 750 genera and more than 19,500 species, traditionally divided into three subfamilies, of which Mimosoideae and Papilionoideae comprises of many nodulating genera in contrast to Caesalpinioideae with few nodulating genera [1, 2]. As per the recent classification of legumes the six subfamilies proposed are Caesalpinioideae, Cercidoideae, Detarioideae, Dialioideae, Duparquetioideae, and Papilionoideae [3]. The species within four (Cercidoideae, Detarioideae, Dialioideae, and Duparquetioideae) newly described sub-families are nonnodulating. The members of the old Mimosoideae have been nested in mimosoid clade within the redefined Caesalpinioideae which now comprises nodulating genera in tribes Ingeae, Mimoseae, Caesalpiniae, and Cassiae [4–6]. Legumes

Kapuganti Jagadis Gupta (ed.), *Nitrogen Metabolism in Plants: Methods and Protocols*, Methods in Molecular Biology, vol. 2057, https://doi.org/10.1007/978-1-4939-9790-9_12,

are cultivated since ancient times for food-feed, shelter, medicines, and livelihood. Earlier researches on legume-rhizobia symbiosis were restricted to few agriculturally important food legume crops and associated species of rhizobia. Since the beginning of exploration of native and wild legumes from various geographical and agroclimatic regions all over the world the number of rhizobial genera has increased to 18 with more than 250 validly defined species (List of Prokaryotic Names with Standing in Nomenclature (LPSN); http://www.bacterio.net) [7–9]. Nowadays, emphasis has been given to research that may greatly help in reducing chemical fertilizers and facilitate the agricultural industry by providing a basis for identifying the best host–rhizobia combinations [10]. The root and stem nodulating rhizobia are classified into 18 genera of which 16 belongs to class alpha-proteobacteria and two to beta-proteobacteria [9]. Also, few unconfirmed gamma-proteobacteria have been reported from root nodules of legumes [7]. The initial approach used for recognition of RNB was dependent mainly on its ability to nodulate legumes and biased toward agronomically important hosts [11]. Later, the advancement in polyphasic approaches toward bacterial taxonomy provided a better insight of the diversity and phylogenetic relationships among RNBs. Presently, along with morphological and phenotypic traits; metabolic patterns analyzed through BIOLOG; and molecular characterization including DNA fingerprinting, phylogenies based on 16S rRNA and protein-coding housekeeping genes, whole genome sequencing, in silico DNA–DNA hybridization, and average nucleotide identity (ANI) are contributing in the description of new genera and species of rhizobia [12, 13]. Sequencing of ribosomal genes of several strains could be expensive therefore PCR-based locus-specific RFLP pattern (ARDRA, Amplified Ribosomal DNA Restriction Analysis) and RPO1 primer [14] based DNA fingerprinting is performed for grouping of rhizobial isolates from different sampling sites based on genetic diversity which generally shows high reproducibility and good agreement with partial or complete gene sequencing.

Multi Locus Sequence Analysis (MLSA) involves genotypic characterization of prokaryotes using the sequences of multiple, conserved protein-coding core genes. MLSA based on individual or concatenated sequences of three to four housekeeping (e.g., *recA*, *atpD*, *glnII*, and *dnaK*) genes can provide sufficient and accurate genetic information of rhizobia [15, 16]. Unlike the conserved housekeeping genes which are located on chromosome, the plasmid-borne accessory genes are strain-specific, subject to horizontal gene transfer (HGT) and help in adaptation of strain to local environmental conditions. Although, accessory symbiotic genes do not reflect the taxonomic position of a strain, characterization of some symbiotic genes (*nodA*, *nodC*, and *nifH*) supplemented with host range studies is routinely performed for symbiotic characterization of RNB and for screening of symbiovars. Cross-inoculation experiments demonstrate the ability of the test strains to elicit effective nodules and helps in determining their host range.

In India studies on nodulation, molecular and symbiotic characterization of the RNB have been carried out from more than decade at BNF and Microbial Genomics Laboratory, Department of Botany, Center of Advanced Study, Jai Narain Vyas University, Jodhpur and the methodology for exploration, identification and characterization of rhizobial diversity associated with various unexplored wild legumes of Indian Thar Desert such as species of *Vachellia*, *Senegalia*, *Mimosa*, *Prosopis*, *Tephrosia*, *Rhynchosia*, *Crotalaria*, and *Alysicarpus* [17–28] have been standardized. Using similar methodology the microsymbiont(s) of less explored native legumes from the biodiversity hotspot in North-Eastern region (Meghalaya) of India have been characterized and novel N-fixing strains of *Bradyrhizobium* are reported [29]. The identification and characterization of indigenous rhizobia associated with various wild and crop legumes provides the basic information of the indigenous rhizobial diversity that has coevolved with host legumes in particular land area and this can be used in future for expanding the cultivated area of legume cash crops. Effective N-fixing bacterial strains with broad host range need to be identified. This will help in developing consortium of promiscuous RNB strains that can be used for preparation of rhizobial inoculums. Cross-inoculation studies provide a base and helps in introducing new crop legumes compatible with native rhizobia in particular area. Here, we explain the general methods for excavation of legume plants with nodules, isolation of rhizobial strains and their preservation, study of some phenotypic traits, and molecular characterization using DNA fingerprinting, phylogenetic analysis based on housekeeping and symbiotic gene sequences. Further, the procedure for tagging of rhizobia with green fluorescent protein (GFP) is explained that facilitates host authentication and cross-inoculation studies by localization of bacteroids. With the help of pipeline of experiments mentioned in this chapter and comprehensive methods described in books such as "The handbook for rhizobia: methods in legume rhizobia technology" by Somasegaran and Hoben [30] and "Working with rhizobia" by Howieson and Dilworth [31] the N-fixing RNB strains can be characterized.

2 Materials

2.1 Collection of Nodules

1. Spade for excavation of plants with nodules from field.
2. Plastic bags for collection of rhizospheric soil.
3. GPS device for recording coordinates of the sampling site.
4. Activated silica beads for nodule storage.

5. 1% glutaraldehyde prepared in 0.4 M phosphate buffer for fixation of nodules for anatomical/microscopic studies.
6. Small pots of approximately 10 × 10 cm or small polybags for rhizobia trapping experiment specifically for tree species.
7. Sterilizing agents [90% ethanol, 0.1% Bavistin^R (fungicide) and 1% sodium hypochlorite] for seed sterilization, moist filter paper or 1% water agar, germinated seedlings of desired legume, sterile sand, sterile distilled water (DW) and electronic weighing balance.

2.2 Isolation and Preservation of Rhizobia

1. Surface sterilizing solutions including 90% ethanol, 1% sodium hypochlorite and 0.1% Bavistin^R. Muslin cloth or tea strainers for avoiding the direct contact of nodules with the sterilizing solutions.
2. Petri plates with Yeast Extract Mannitol Agar (YEMA) media with Congo red (CR) dye, inoculating loops, autoclaved forceps, DW and parafilm for sealing.
3. The CR dye is used as an indicator for differentiation of rhizobia (do not absorb the dye) with other nonrhizobial endophytic bacteria. Stock solution of dye is prepared by adding 0.25 g/100 ml of DW. In CR-YEMA medium the final concentration of CR is adjusted to 25 μg ml^{-1} [30].
4. YEMA media: Add 10 g Mannitol, 0.2 g $MgSO_4{\cdot}7H_2O$, 0.5 g K_2HPO_4, 0.1 g NaCl, 0.4 g Yeast Extract, 15 g agar-agar in DW and make volume up to 1 l. Set pH at 6.8.
5. Tryptone Yeast (TY) media: Add 5 g Tryptone, 3 g Yeast Extract, 0.87 g $CaCl_2{\cdot}6H_2O$, 15 g agar-agar in DW and make volume up to 1 l. Adjust pH at 6.8–7.0.
6. Agar slants/stabs prepared of preferred growth medium (YEMA or TY).
7. Cryopreservation: Deep freezers, 80% glycerol, activated broth of pure rhizobial culture, 0.89% saline/TY broth, 2 ml sterile cryo-tubes and vortex.
8. Lyophilization: 1:1 mixture of 10% (w/v) peptone and 10% (w/v) glutamate recommended specially for rhizobial cultures, cotton wool, broth suspension of rhizobia, high quality glass ampoules, Pasteur pipette, lyophilizer machine, etc.

2.3 Phenotypic Characterization of Rhizobia

1. Activated pure broth cultures of rhizobia and 0.89% saline.
2. YEMA media with different concentrations (0.5%, 0.75%, 1%, 2%, and 3%) of NaCl salt.
3. Different buffers for adjusting at a desired range of pH: HOMOPIPES (Homopiperazine-*N*, *N'*-*bis*-2-ethanesulfonic acid) for pH 4.0–5.0; MES (2-[*N*-morpholino]-ethanesulfonic acid) for pH 5.5, 6.0; HEPES (4-[2-Hydroxyethyl] piperazine-

1-ethanesulfonic acid) for pH 7.0, 8.0, 8.5; CHES (*N*-Cyclohexyl-2-aminoethanesulfonic acid) for pH 9.0, 9.5, 10.0 [32]; 1 N HCl and 1 N NaOH for adjusting pH.

4. Different HiMedia antibiotic discs: chloramphenicol (20 μg ml^{-1}), gentamycin (40 μg ml^{-1}), kanamycin (50 μg ml^{-1}), nalidixic acid (75 μg ml^{-1}), spectinomycin (100 μg ml^{-1}), and streptomycin (100 μg ml^{-1}) used for rhizobia [30], cotton swabs and scale for measurement of zones of inhibition.
5. Andrade's peptone water: Add 10 g Peptone, 5 g NaCl, and 5 ml Andrade's indicator to 1 l of DW and adjust pH to 7.4. Prepare Andrade's indicator by adding 0.1 g Acid fuschin to 16 ml 1 N NaOH solution and make volume up to 100 ml with DW. 24-well plates, HiMedia discs of different sugars to be tested (e.g., adonitol, arabinose, cellobiose, dextrose, dulcitol, galactose, fructose, inositol, inulin, lactose, maltose, mannitol, mannose, melibiose, raffinose, rhamnose, salicin, sorbitol, sucrose, trehalose, and xylose).

2.4 Molecular Characterization of Rhizobia

1. For DNA isolation: activated TY broth, Tris-saturated phenol, chloroform, RNase A (10 mg ml^{-1}), 1.5 ml microcentrifuge tubes, STE (Sodium Chloride Tris–EDTA) buffer (100 mM NaCl, 10 mM Tris–HCl, 1 mM EDTA and adjust pH to 8.0), and TE (Tris–EDTA) buffer (10 mM Tris–HCl, 1 mM EDTA and adjust pH to 8.0).
2. Cell template preparation: TY agar media and 0.89% saline.
3. PCR reactions: 0.2 ml PCR tubes, 10× Taq buffer, 25 mM $MgCl_2$, 10 mM dNTPs mix (2.5 mM each), 3 Uμl^{-1} *Taq* DNA polymerase, dimethyl sulfoxide (DMSO), nuclease-free water, and the primers listed in Tables 1 and 2 for amplification of various gene fragments.
4. For ARDRA (Amplified rDNA restriction analysis), amplified products of 16S rDNA, tetracutter restriction enzymes such as *Msp*I or *Alu*I (10 Uμl^{-1}), 10× buffer and nuclease free water are required.
5. For sequencing reactions: PCR product purification kit, 3.2 pmol of terminal and internal primers, Applied Biosystems Big Dye terminator and 5× sequencing buffer, 30–40 ng of purified PCR product, nuclease free water, 125 mM EDTA, 3 M sodium acetate, 80% ethanol, 70% ethanol, and Kimwipes.
6. For gel electrophoresis: Mini or midi electrophoresis system, agarose, SYBR Safe or EtBr, 6× gel-loading buffer, DNA ladder, and 1× TAE buffer (4.85 g l^{-1} Tris, 2 ml l^{-1} of 0.5 M EDTA [pH 8.0], and 1.142 ml l^{-1} glacial acetic acid in deionized water).
7. Bioinformatics software(s) for analysis of DNA sequences and phylogenetic tree reconstruction.

Table 1
List of primers for amplification and sequencing of housekeeping genes in rhizobia

Primer	Oligonucleotide sequence[a] (5′ → 3′)	Reference	Rhizobia type	Amplicon product name and its size
18F	AGAGTTTGATCCTGGCTCAG	[33]	*Ensifer*, *Bradyrhizobium*	16S ribosomal RNA (1500 bp)
1492R	CTACGGCTACCTTGTTACG			
fD1	AGAGTTTGATCCTGGCTCAG		*Bradyrhizobium*	
rD1	CTTAAGGAGGTGATCCAGCC			
800F	GTAGTCCACGCCGTAAACGA	[34]	Internal primers	
820R	CATCGTTTACGGCGTGGACT			
atpD294F	ATCGGCGAGCCGGTCGACGA	[35]	*Bradyrhizobium*	ATP synthase F1 beta subunit (500 bp)
atpD771R	GCCGACACTTCCGAACCNGCCTG			
atpD352F	GGCCGCATCATSAACGTCATC	[16]	*Ensifer*	
atpD871R	AGAGCCGACACTTCMGARCC			
TSglnIIF	AAGCTCGAGTACATCTGGCTCGACGG	[15]	*Bradyrhizobium*	Glutamine synthetase II (620 bp)
TSglnIIR	GAGCCGTTCCAGTCGGTGTCG			
GSII-1	AACGCAGATCAAGGAATTCG	[36]	*Ensifer*	Glutamine synthetase II (600 bp)
GSII-2	ATGCCCGAGCCGTTCCAGTC			
TSrecAF	CAACTGCNYTGCGTATCGTCGAAGG	[15]	*Bradyrhizobium*	Recombination protein (600 bp)
TSrecAR	CGGATCTGGTTGATGAAGATCACCATG			
recA6F	CGKCTSGTAGAGGAYAAATCGGTGGA	[35]	*Ensifer*	Recombination protein (550 bp)
recA555R	CGRATCTGGTTGATGAAGATCACCAT			
TSdnaK2	GTACATGGCCTCGCCGAGCTTCA	[37]	*Ensifer*, *Bradyrhizobium*	Molecular chaperone (300 bp)
TSdnaK3	AAGGAGCAGCAGATCCGCATCCA			

[a]N = A, C, G or T; R = A, G; Y = C or T; K = G or T; and S = C or G

Table 2
List of primers for amplification and sequencing of symbiotic genes in rhizobia

Primer	Oligonucleotide sequence[a] (5′ → 3′)	Reference	Rhizobia type	Amplicon product name and its size
nodAf.brad	GTYCAGTGGAGSSTKCGCTGGG	[38]	*Bradyrhizobium*	*N*-acyltransferase nodulation protein (550 bp)
nodAr.brad	TCACARCTCKGGCCCGTTCCG			
nodA1	TGCRGTGGAARNTRNNCTGGGAAA	[39]	*Ensifer*, *Bradyrhizobium*	*N*-acyltransferase nodulation protein (650 bp)
nodA2	GGNCCGTCRTCRAAWGTCARGTA			
nodA3	TCATAGCTCYGRACCGTTCCG	[40]		
nifHF	TACGGNAARGGSGGNATCGGCAA	[41]	*Ensifer*, *Bradyrhizobium*	Nitrogenase Fe protein (750 bp)
nifHI	AGCATGTCYTCSAGYTCNTCCA			
nodCF	AYGTHGTYGAYGACGGTTC			*N*-acetylglucosaminyl transferase (950 bp)
nodCI	CGYGACAGCCANTCKCTATTG			
nifH1	CGTTTTACGGCAAGGGCGGTATCGGCA	[42]	*Ensifer*	Nitrogenase Fe protein (780 bp)
nifH2	TCCTCCAGCTCCTCCATGGTGATCGG			

[a]N = A, C, G or T; R = A, G; W = A, T; Y=C or T; H = A, C or T; K = G or T; and S = C or G

2.5 Authentication and Cross Inoculation Studies

1. Autoclavable draining plastic pots (15 × 11 cm), polyvinyl chloride (PVC) pipes with rubber lids, small marbles/plastic beads, mixture of soil and river sand or vermiculite/soilrite and river sand.
2. Healthy viable seeds of legumes, surface sterilizing solutions (90% ethanol, 1% sodium hypochlorite, 0.1% Bavistin^{R}), autoclaved DW, sterile plates containing sterilized moist filter paper or 1% water agar for seed germination.
3. To prepare stocks 1–6 for nitrogen free nutrient solution [31, 43] the chemicals are listed in Table 3. First prepare solution A as described in Table 3, then prepare solution B (N- solution) by adding 200 ml of stock A + 400 ml of stock 5 + 1400 ml of DW, adjust pH to 6.5. For preparing starter solution add 0.5 g l^{-1} KNO_3 to solution B.
4. Pure RNB strains activated on TY agar plates and 1% (w/v) sucrose solution. CR-YEMA plates and other materials for reisolation of rhizobia from root nodules. Materials for RPO1 DNA fingerprinting to compare the parental and reisolate strains.

2.6 GFP Tagging

1. TY agar and broth, Luria broth (Add 10 g Tryptone, 5 g Yeast Extract, and 10 g NaCl to 1 l DW), 10 mM $MgSO_4$ solution, and L-shaped spreaders.

Table 3
Chemical composition for nitrogen-free nutrient solution

Stock 1	$MgSO_4 \cdot 7H_2O$	24.6 g/500 ml
Stock 2	KH_2PO_4	13.6 g/500 ml
Stock 3	K_2SO_4	35.0 g/500 ml
Stock 4	Sodium ferric EDTA	5.0 g/500 ml
Stock 5	$CaSO_4 \cdot 2H_2O$	2.49 g/2 l
Stock 6	Trace elements H_3BO_3 $Na_2MoO_4 \cdot 2H_2O$ $ZnSO_4 \cdot 7H_2O$ $MnSO_4 \cdot 5H_2O$ $CoSO_4 \cdot 6H_2O$ $CuSO_4 \cdot 5H_2O$	Composition for 100 ml (store at 4 °C) 0.046 g 0.001 g 0.053 g 0.004 g 0.014 g 0.012 g
Solution A = 500 ml stock 1 + 500 ml stock 2+ 500 ml stock 3 + 500 ml stock 4 = 2000 ml + 4 ml of trace stock 6		
Solution B (nitrogen free nutrient solution) = 200 ml of solution A + 400 ml of stock 5 + 1400 ml DW		
Starter solution = solution B + 0.5 g/l KNO_3 (pH after autoclaving is 6.5)		

2. *E. coli* strains with donor plasmid pHC60 (containing *gfp* gene) [44] and helper plasmid pRK2013 [45], and the recipient rhizobial strains.
3. Preparation of antibiotics stocks: Streptomycin- 100 mg in 1 ml of autoclaved DW; Tetracycline- 10 mg in 1 ml of 80% ethanol; Kanamycin- 50 mg in 1 ml of autoclaved DW. Antibiotics stocks are filter sterilized (0.2 μm filter) and stored at −20 °C. Tetracycline is light sensitive therefore the vial should be covered with foil.
4. RDM stock A solution: Add 6 g KNO_3, 1 g $CaCl_2 \cdot 2H_2O$, and 2.5 g $MgSO_4 \cdot 7H_2O$ in 1 l DW. RDM stock B solution: Add 10 g K_2HPO_4, 10 g KH_2PO_4, 0.1 g $FeCl_3 \cdot 6H_2O$ in 1 l DW.
5. Biotin stock (0.25 mg ml^{-1}) solution: Add 0.025 g in 100 ml autoclaved deionized water, mix the solution properly and filter-sterilize. Store the vial at 4 °C wrapped in foil.
6. Thiamine stock (10 mg ml^{-1}) solution: Add 0.5 g in 50 ml autoclaved deionized water and filter sterilize. Store the vial at 4 °C wrapped in foil.
7. Rhizobium Defined medium (RDM): For 500 ml media add 50 ml RDM stock A, 50 ml RDM stock B, 2.5 g sucrose, 7.5 g agar-agar and make up the volume with DW. Autoclave at 121 °C for 20 min, and cool to approximately 55 °C then add 2 ml of biotin stock, 500 μl of thiamine stock, 500 μl each of tetracycline and streptomycin stocks.

3 Methods

3.1 Field Survey, Rhizobia-Trapping Experiment, and Nodule Hunting

1. Prepare the list of native legumes of the area/region to be explored. Do extensive surveys for nodulating native/wild legumes and identify plants with the help of local flora and taxonomist. Dig out healthy plant with a spade after making a circle of approximately 10 cm radius and 30 cm depth around the plant (Fig. 1a). Place the excavated plants in a sieve and flush under the gentle stream of water to remove all the adherent soil particles. Record the GPS coordinates of the sampling site, data related to number of nodules per plant (Fig. 1b) and type of nodules (Fig. 1c, d). Prepare the herbarium sheets (both digital and hard copy) of each species and deposit them in herbarium to obtain accession numbers.
2. Collect 10–15 nodules per plant for isolation of rhizobia. Air dry few nodules and then transfer in vials containing activated silica beads for storage, and also fix some nodules in vials containing 1% glutaraldehyde for anatomical/microscopic studies.

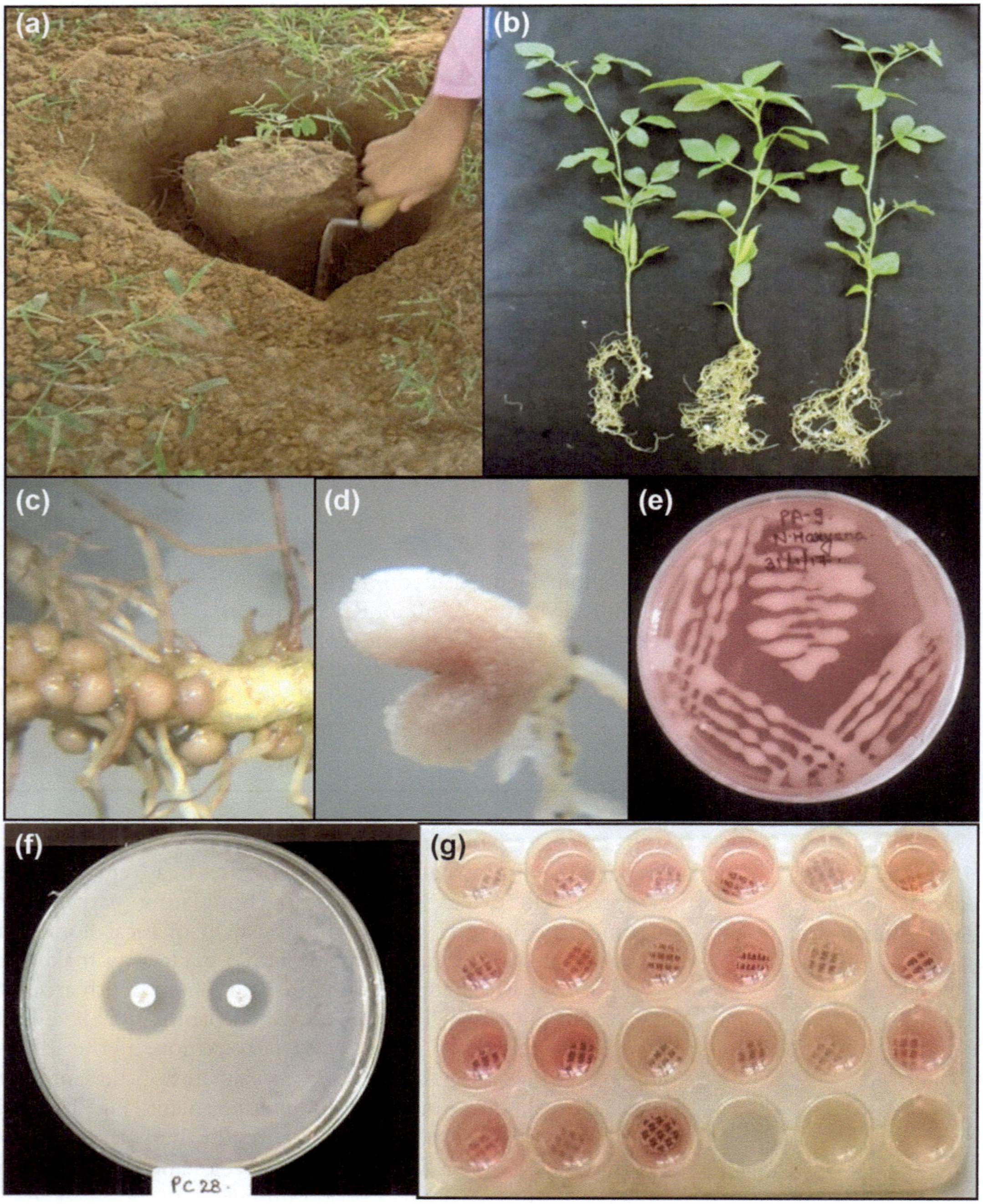

Fig. 1 Excavation of *Tephrosia purpurea* plantlet from field (**a**), excavated plants of *Cyamopsis tetragonoloba* with nodules attached to roots (**b**), determinate nodules of *Aeschynomene americana* (**c**), indeterminate nodule of *Trigonella foenum-graecum* (**d**), plate showing growth of *Ensifer* strain PA9 on CR-YEMA media (**e**), intrinsic antibiotic resistance (**f**), and carbon utilization patterns of RNB strains using HiMedia discs (**g**)

3. For trapping experiment setup collect the rhizospheric soil and fill it in pots/polybags with proper labeling (*see* **Note 1**). Seeds of wild/native legumes often possess seed dormancy. Hot water treatment, acid treatment or scarification of seeds with sandpaper is needed to break the dormancy. In laboratory, surface sterilize the seeds of legume with 90% ethanol for 1 min, 0.1% Bavistin^{R} for 1 min followed by two times washing with DW, then transfer seeds in 1% sodium hypochlorite for 6 min followed by six times washing with DW. After giving hot water treatment to seeds for 15 min, place them on a moist filter paper or 1% water-agar and keep in dark within the seed germinator maintained at 28 °C.
4. The germinated seedlings (after 2–4 days) are transferred in pot/polybags filled with rhizospheric soil. Keep the pots under natural environmental conditions in the net house/poly house and regularly water them with sterile water. After 8–10 weeks, excavate the plants for their nodules; record the number of nodules per plant; root and shoot length; and fresh and dry weight/biomass of the plants.

3.2 Isolation and Preservation of Root Nodule Bacteria

1. Properly wash the excavated root nodules with sterile DW. Immerse the dried (in case of nodules preserved in silica gel) nodules in sterile DW for 3–4 h for rehydration.
2. For surface sterilization, the nodules are either tied in small pieces of muslin cloth or placed in tea strainers and rinsed for 1 min in 90% ethanol, 1 min in 0.1% Bavistin^{R} followed by two times washing with sterile DW, then placed in 1% sodium hypochlorite for 6 min followed by six times washing with sterile DW.
3. After surface sterilization the individual nodules are placed in an aseptic watch glass, cut into two halves with sterile scalpel and one drop of sterile DW is added to each section of nodule. Streak the white exudate from each nodule on the CR-YEMA petri plates, and incubate the plates at 28 °C. Observe the plates for growth of colonies. Pick up the raised, entire, circular, white, translucent or opaque, exopolysaccharide secreting rhizobia-like colonies (*see* **Note 2**). Perform quadrant streaking on CR-YEMA media for purification of colonies from the master plate. Record the duration of growth and morphological characteristics of pure strains (Fig. 1e).
4. For short term preservation the pure RNB strains are regularly sub-cultured and preserved on agar slants/stabs kept at low temperature (4 °C).
5. For long term storage the cultures are preserved as glycerol stocks in deep freezer at −20 °C or −80 °C. To 1.7 ml of 0.89% saline/TY broth 300 μl of 80% glycerol is added in a 2 ml cryo-

tube. To this mixed solution 3–4 loops full of pure fresh culture is added, vortexed and stored at −20 °C or −80 °C.

6. Lyophilization: As described by Howieson and Dilworth [31] for long term storage (by freeze-drying) the purified cultures are first suspended in a 1:1 solution of 10% (w/v) peptone and 10% (w/v) glutamate. The ampoules are prepared by placing the sterile cotton at its bottom, the open end of ampoule is loosely plugged and autoclaved at 121 °C for 30 min. Few drops of the bacterial suspension is added to sterile cotton wool placed at the bottom of sterile ampoules. After labeling the ampoules are placed under vacuum for dehydration following the manufacturer's instructions. After proper freeze-drying the ampoules are sealed and stored at 4 °C. For each strain about 6–8 ampoules are prepared.

3.3 Study of Selective Phenotypic Traits of Root Nodule Bacteria

1. Salt tolerance: Prepare dilutions of activated broth culture concentrated at $OD_{600} = 1$ in 0.89% saline. Spot inoculate 10 μl of this bacterial suspension on YEMA media buffered with 20 mM HEPES and supplemented with different concentrations (0.5%, 0.75%, 1%, 2% and 3%) of salt (NaCl). Spot inoculated plates are incubated at 28 °C for 5–10 days and the growth of rhizobial strains at different concentrations of NaCl is recorded.
2. pH tolerance range: Spot inoculate 10 μl of activated broth suspension (as mentioned earlier) on YEMA media adjusted at different pH with buffers as described in Subheading 2.3. Incubate the plates at 28 °C for 5–10 days and observe the growth of rhizobial strains at different pH.
3. Intrinsic Antibiotic Resistance (IAR) pattern: it is tested according to Kirby Boyer's Disc diffusion method [46]. Swab the individual rhizobial strains on petriplates containing YEMA media and aseptically place HiMedia antibiotic discs with different concentration mentioned in Subheading 2.3. Incubate plates for 2 days at 28 + 2 °C. After incubation record the zone of inhibition in mm (Fig. 1f).
4. Carbon utilization pattern: In sterile 24 well plates add individual HiMedia sugar discs and Andrade's peptone water to each well, inoculate each well with activated broth culture. Keep the plates at 28 °C and observe after every 24 h. Metabolism of particular sugar by the strain will release acids thus, lowering the pH of media and changing the color of medium, from pale straw color to pink color (Fig. 1g).

3.4 Genotypic Characterization of Root Nodule Bacteria

1. For bacterial DNA isolation as described by Cheng and Jiang [47] take 1 ml of activated bacterial broth in 1.5 ml microcentrifuge tube. Centrifuge at 9000 × *g* for 5 min at room temperature. Remove the supernatant and wash the pellet with

STE buffer twice by vortexing and centrifugation at 9000 × *g* for 5 min. Resuspend the pellet in 200 μl TE buffer and add 100 μl saturated phenol to it. Vortex to lyse the cells. Centrifuge at 15,000 × *g* for 5 min at 4 °C to separate the aqueous phase from the organic phase. Take 155 μl of upper aqueous phase and transfer in a new vial. To this add 40 μl TE buffer and 5 μl RNase A, incubate at 37 °C for 15 min. Now after incubation add 100 μl of chloroform to it and invert the vials upside down to mix the components well, then centrifuge at 15,000 × *g* for 5 min at 4 °C. Transfer 160 μl of upper aqueous phase in a new vial and dilute with 40 μl TE buffer. Determine the concentration (ng/μl), 260/230 and 260/280 ratio using nanodrop spectrophotometer and run the pure DNA on a 0.8% (w/v) agarose gel for half an hour.

2. Cell template preparation: Activate pure cultures of RNB strains on TY agar medium. Scrap four to five loops full of cells from freshly activated plates in 1 ml of 0.89% (w/v) sterile saline solution. Vortex the suspension and centrifuge at 9000 × *g* for about 10 min. Discard the supernatant and resuspend the pellet in 1 ml saline (*see* **Note 3**). One hundred microliter of thoroughly vortexed suspension is taken in a new vial and to it 900 μl of saline is added. After vortexing its absorbance/optical density is taken at 600 nm. Total OD is calculated from dilute OD and is used to make final required volume. One hundred microliter of saline of OD 2 and OD 6 are prepared for gene amplification and DNA fingerprinting respectively. Store the templates at 4 °C.
3. Agarose gel preparation: agarose gel prepared in 1× TAE buffer is casted to a thickness of 8 mm. Gels are prestained by incorporating SYBR Safe (8 μl/100 ml) or EtBr (3 μl/100 ml) in molten agarose. For genomic DNA and 16S rDNA PCR products, less concentrated gel, that is, 0.8% (w/v), for other housekeeping and *sym* genes 1.5% (w/v) gel and for DNA fingerprinting 2% (w/v) gel is required.
4. DNA fingerprinting (RAPD-randomly amplified polymorphic DNA) using RPO1 primer: single *nif*-directed RPO1 primer (5′AATTTTCAAGCGTCGTGCCA3′) described by Richardson et al. [14] is used for DNA fingerprinting. Prepare 20 μl reaction mixture containing 1 μl of cell template ($OD_{600} = 6$) or DNA sample diluted to 50–100 ng/μl, 2 μl of 10× Taq buffer, 3 μl of 25 mM $MgCl_2$, 1.2 μl of dNTP mix (2.5 mM each), 1 μl of DMSO, 1.2 μl of 50 μM RPO1 primer, 0.35 μl of 3 $U\mu l^{-1}$ *Taq* DNA polymerase and nuclease free water to make its volume up to 20 μl. Set up the thermal cycler with cycling conditions as follows: 5 min 94 °C, 5 × (30 s 94 °C, 1 min 55 °C, 1.5 min 72 °C), 30 × (30 s 94 °C, 25 s 58 °C, 30 s 72 °C), followed by final extension of 7 min at 72 °C. Load the

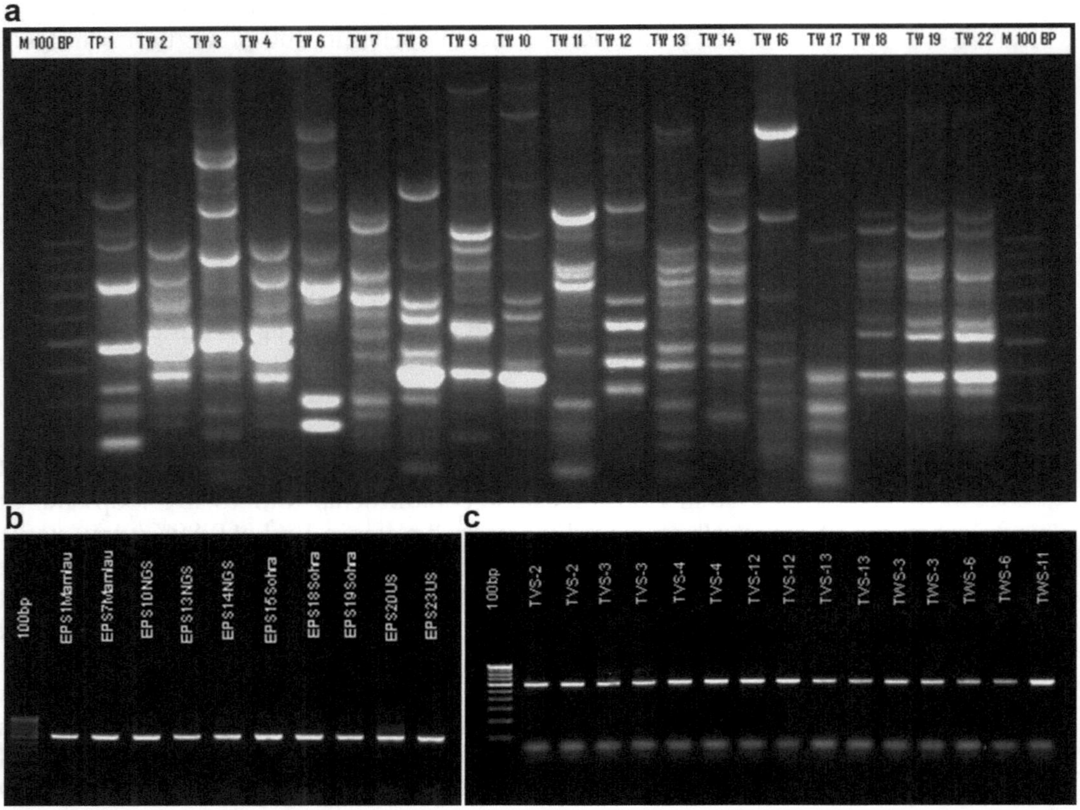

Fig. 2 Gel showing RPO1 DNA fingerprinting pattern of RNB strains isolated from *Tephrosia* sp. (**a**), gel with approximately 600 bp band of *recA* (**b**), and 550 bp band of *nodA* (**c**) gene amplified in *Bradyrhizobium* strains

PCR products (20 μl) and DNA ladder in wells of a 2% (w/v) gel and run at 80 V for 1–2 h. Observe the banding patterns for grouping of strains on the basis of genetic fingerprints (Fig. 2a).

5. ARDRA (**A**mplified **R**ibosomal **D**NA **R**estriction **A**nalysis): Prepare a reaction mixture by adding 10 μl of 16S rDNA PCR product, 2 μl of 10× buffer, 0.5 μl of *Alu*I or *Msp*I enzyme (10 $U\mu l^{-1}$) and 7.5 μl of nuclease free water. Keep the mix at 37 °C overnight in incubator and run the digested product along with DNA ladder on a 2% (w/v) gel for 1–2 h. at 80 V. Observe the banding patterns using gel documentation system for analysis of genetic fingerprints.
6. PCR amplification of housekeeping and symbiotic genes: Prepare reaction mixture of 25 μl volume containing 1.25 μl sample DNA, 2.5 μl 10× Taq buffer, volume of 25 mM $MgCl_2$ depending on the type of gene fragment amplified (Tables 4 and 5), 1.5 μl of dNTP mix (2.5 mM each), 1.25 μl of DMSO, 0.6 μl each of forward and reverse primer, 0.25 μl of 3 $U\mu l^{-1}$

Table 4
Thermal cycling conditions for amplification of housekeeping genes in rhizobia

Primer pair	$MgCl_2$ final conc. (mM)	PCR cycling conditions
fD1 and rD1	1.5–1.8	5 min 95 °C, 30 × (1 min 94 °C, 1 min 55 °C, 2 min 72 °C), 7 min 72 °C
18F and 1492R	2.0	5 min 95 °C, 35 × (1 min 94 °C, 1 min 53 °C, 1 min 72 °C), 7 min 72 °C
atpD352F and atpD871R	1.75–2.0	5 min 95 °C, 3 × (2 min 94 °C, 2 min 64 °C, 1 min 72 °C), 30 × (30 s 94 °C, 1 min 64 °C, 1 min 72 °C) 5 min 72 °C
atpD294F and atpD771R	2.0	5 min 94 °C, 35 × (45 s 95 °C, 30 s 58 °C, 1.5 min 72 °C), 7 min 72 °C
recA6F and recA555R	1.0–2.0	5 min 94 °C, 30 × (45 s 94 °C, 1 min 60 °C, 1.5 min 72 °C), 7 min 72 °C
TSrecAF and TSrecAR	1.0–2.0	5 min 94 °C, 30 × (45 s 94 °C, 1 min 57 °C, 1.5 min 72 °C), 7 min 72 °C
TSglnIIF and TSglnIIR	2.0	5 min 95 °C, 3 × (2 min 94 °C, 2 min 60 °C, 1 min 72 °C), 30 × (30 s 94 °C, 1 min 62 °C, 1 min 72 °C), 5 min 72 °C
GSII-1 and GSII-2	1.5–2.0	5 min 95 °C, 3 × (2 min 94 °C, 2 min 50 °C, 1 min 72 °C), 30 × (30 s 94 °C, 1 min 54 °C, 1 min 72 °C), 5 min 72 °C
TSdnaK2 and TSdnaK3	1.5–2.0	5 min 95 °C, 35 × (30 s 94 °C, 30 s 62 °C, 45 s 72 °C), 5 min 72 °C

Table 5
Thermal cycling conditions for amplification of symbiotic genes in rhizobia

Primer pair	$MgCl_2$ final conc. (mM)	PCR cycling conditions
nodA1 and nodA2/nodA3	2.0	5 min 94 °C, 5 × (30 s 94 °C, 30 s 55 °C, 1 min 72 °C), 30 × (30 s 94 °C, 45 s 62 °C, 1.5 min 72 °C), 7 min 72 °C
nodAf.brad and nodAr.brad	1.75–2.0	5 min 95 °C, 35 × (1 min 94 °C, 45 s 57 °C, 1 min 72 °C), 10 min 72 °C
nifHF and nifHI	2.0	5 min 94 °C, 25 × (30 s 94 °C, 30 s 57 °C, 30 s 72 °C), 7 min 72 °C
nifH1 and nifH2	2.0	30 s 95 °C, 5 × (30 s 94 °C, 30 s 55 °C, 1 min 72 °C), 30 × (30 s 94 °C, 30 s 65 °C, 1 min 72 °C), 5 min 72 °C
nodCF and nodCI	2.0	3 min 95 °C, 35 × (1 min 94 °C, 1 min 55 °C, 2 min 72 °C), 3 min 72 °C

Taq DNA polymerase, and nuclease-free water according to the reaction mixture to make its volume up to 25 μl. Set and save the different thermal cycling protocols in the PCR, as described in Tables 4 and 5 corresponding to various genes. After cycling, load the PCR products along with DNA ladder on a 1.5% (w/v) gel at 80 V for 30 min. The band size is confirmed (Fig. 2b, c) before proceeding for more reactions to pool up the amplicons for sequencing.

7. PCR product purification and sequencing: Purify the PCR products to remove primers, dNTPs, *Taq* DNA polymerase and primer dimers using the PCR product purification kit as per the manufacturer's instructions. For carrying out sequencing-PCR prepare a quarter reaction mix of 10 μl volume by adding 2 μl of Big Dye Terminator, 1 μl of 5× sequencing buffer, 4 μl of 30–40 ng of purified PCR product, 1 μl of 3.2 pmol of primer, and 2 μl of nuclease-free water. Proceed for sequencing-PCR with following cycling conditions: 2 min 96 °C, 25 × (10 s 96 °C, 5 s 50 °C, 4 min 60 °C) and store at 14 °C. For ethanol/EDTA precipitation method, add 1 μl of 125 mM EDTA, 1 μl of 3 M sodium acetate (pH 5.2), and 25 μl of 80% ethanol to 10 μl of amplified sequencing-PCR product. Thoroughly mix the components with pipette and keep at room temperature for 20 min. Centrifuge the reaction mix at 18,000 × *g* for 30 min. Remove supernatant with the pipette and immediately add 125 μl of freshly prepared 70% ethanol. Centrifuge at 18,000 × *g* for 5 min at room temperature. Completely remove the supernatant with the help of Kim wipes and store at −20 °C when completely dried (*see* **Note 4**). After ethanol precipitation all DNA sequencing steps are carried out as described by the manufacturer (Applied Biosystems) by using a dye terminator cycle sequencing ready reaction kit and automated sequencer (ABI Model).
8. Sequence analysis and editing: Observe the chromatograms, edit and assemble the sequences using BioEdit software (http://www.mbio.ncsu.edu/BioEdit/bioedit.html) or GeneTool Lite (version1.0 Double twist Inc., Oakland, CA, USA) (Fig. 3a). The NCBI (National Center for Biotechnology Information) nucleotide BLAST (Basic Local Alignment Search Tool, https://blast.ncbi.nlm.nih.gov/) online tool can be used to identify and search similar sequences in the NCBI's nucleotide sequence database. The assembled sequences can be deposited in the GenBank database using the NCBI Sequin software (https://www.ncbi.nlm.nih.gov/Sequin/). To obtain the GenBank accession numbers Sequin files prepared for individual gene sequences must be mailed to gb-sub@ncbi.nlm.nih.gov.

3.5 Phylogenetic Analysis

1. For constructing phylogenetic trees the most widely used and freely available software, MEGA (Molecular Evolutionary Genetics Analysis) version 7 (http://www.megasoftware.net/) or the latest version available can be used [48].
2. For creating a new alignment in MEGA click on "align" and then on "edit/build alignment." Select the option to "create a

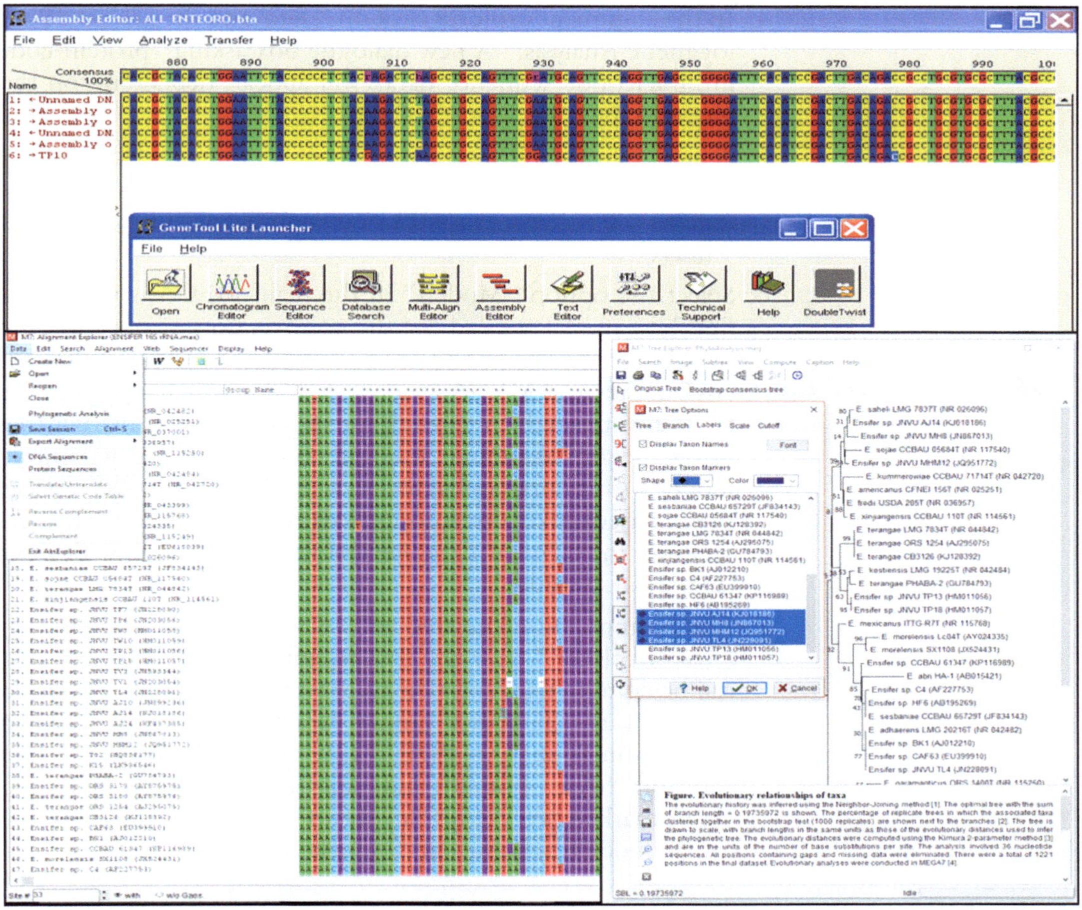

Fig. 3 GeneTool Lite software used for editing and assembly of DNA sequences (**a**), Alignment Explorer (**b**), and Tree Explorer (**c**) of MEGA (version 7) software used for phylogenetic analysis

new alignment" and 'DNA' in the next dialogue box for building a DNA sequence alignment. The alignment explorer will open, click on insert a new blank sequence or just press 'Ctrl+N' to add the sequences (in FASTA format) to be aligned.

3. Download the sequences in FASTA format for all type/reference strains and close relatives from the NCBI GenBank (https://www.ncbi.nlm.nih.gov/nucleotide) and create "mas" files for different genes separately by pasting the sequences one by one in the alignment explorer. Edit the names of species in the Species/Abbrv column and sort them alphabetically. Go to "Edit" option and select all the unaligned sequences and then click on "align by ClustalW" under "Alignment" tool to align the selected sequences. After pairwise and multiple alignment save the aligned session with the extension '.mas' (Fig. 3b).

4. In the alignment explorer click on 'Data' and then select "Phylogenetic Analysis." A new dialogue box asking "protein-coding nucleotide sequence data" will appear, depending on the sequence type, click on "Yes" or "No."
5. A new "Sequence Data Explorer" file will open, select or deselect the sequences of different strains for construction of phylogenetic tree. Save this MEGA session file with the extension ".mdsx".
6. Select the option "Models" in the main MEGA window and find best model needed for the construction of phylogenetic tree. Once the strains and model is finalized click on "Phylogeny" in the main MEGA window to decide which type of tree is to be constructed. In general, for rhizobia, to reconstruct tree using neighbor-joining method the Kimura 2-parameter model with bootstrap values between 500 and 1000 replications is used and for tree using Maximum Likelihood statistical method, the General Time Reversible model is used with similar bootstrap values.
7. After computation the tree will be seen in a new window of "Tree Explorer." Save the tree session in "MTS format" that can be edited. Also, copy the tree in word file using the option of "Copy to Clipboard" under "Image" or can be saved in other formats such as PDF, TIFF, etc. For editing the tree click on "tree options" and edit labels, change font, add taxon markers, etc. (Fig. 3c). Lower bootstrap values (<50) can be hidden by using the "branch" tab.

3.6 Nodulation Assays and Symbiotaxonomy

1. The RNB strains are confirmed as microsymbiont of their host legume through inoculation/infection in sterile conditions and reisolation of symbiotic bacteria from nodules formed on the infected/treated plants in glasshouse pot experiments/free-draining pot method described by Yates et al. [49]. Select wild as well as crop legumes to be tested for cross-inoculation studies to reveal the host range of rhizobial strains.
2. Autoclave the PVC pipes along with lids and marbles/beads at 121 °C for 45 min. Fill the plastic pots with equal quantity of river sand and soil or mix 1/4th quantity of vermiculite/soilrite and 3/4th quantity of river sand. At the bottom of each pot absorbent paper is placed to prevent sand leakage. Flush the mix filled in pots with hot boiling water for about two to three times to remove the inorganic nitrogen. Cover the well-drained pots with foil and autoclave at 121 °C for about 2–3 h.
3. Meanwhile, for germinating the seeds of selected legumes first scarify them with the help of sandpaper; then surface sterilize with 90% ethanol, 0.1% Bavistin^{R}, followed by two times

washing with sterile DW and then immerse in 1% sodium hypochlorite followed by six times washing with sterile DW. After sterilization the seeds are kept imbibed for 15–20 min and later transferred to sterile petriplates containing filter paper (moistened) or 1% water agar (*see* **Note 5**). Plates are covered with foil, kept in seed germinator at 28 °C and monitored until adequate growth (emergence of radicles) for transplanting.

4. For preparation of inoculums, the RNB strains are streaked on TY agar plates and incubated at 28 °C for 3–5 days, until confluent growth had occurred. After sufficient growth appears the cells are scraped in 1% sucrose solution when the germinated seedlings have to be transplanted into the pots. Six germinated seeds with approximately 1–2 cm radicles are sown in each pot. The germinated seeds are sown to a depth of 5–10 mm using sterile forceps. Hypocotyls of each seedling are inoculated with 1 ml suspension of bacterial strains containing approximately 10^7 cells for infection and nodulation (*see* **Note 6**).
5. The pots for each strain tested on a particular legume host and controls are prepared in triplicates. The un-inoculated controls supplemented with nitrogen source KNO_3 (10 ml of 0.25 M KNO_3 per pot weekly) serve as "N+" controls and the un-inoculated controls without supplements of nitrates serve as "N−" control (Fig. 4a) (*see* **Note 7**).
6. Growth conditions and watering of pots: The plants are maintained at 28−32 °C in a naturally lit glasshouse or polycarbonate growth chambers. After every second day cooled autoclaved water or N free nutrient solution is added to plants. 20 ml of starter solution containing KNO_3 is added to all the pots after sowing for establishment of young seedlings. 20 ml of N free nutrient solution is added weekly to the pots till 3 weeks and then increased to 40 ml thereafter depending upon the growth of plants.
7. After 8 weeks excavate the plants and analyze each treatment for nodulation (Fig. 4b). Note nodule morphology, color, position on the roots, and number per plant for each treatment which has nodulation. Both fresh and dry weight of the plants is recorded. Even the bacteria are reisolated from few nodules of the treated plants and the cell templates are prepared which can be used for RPO1 DNA fingerprinting (as described in Subheading 3.4) to compare the banding patterns of reisolates with the parental strains in order to confirm nodule occupancy.

Fig. 4 Image showing set up for free draining pot-experiment (**a**), nodulation observed in *Vigna unguiculata* plants inoculated with *Ensifer* sp. PC2 and pale yellow N-control plant without nodulation (**b**), GFP-tagged bacterial rods (**c**), and expression of GFP within root nodule (**d**)

3.7 Insertion of GFP Reporter Gene in Rhizobia and Screening of Transformants

1. Inoculate 5 ml TY broth with single colony of recipient rhizobial strain and add 5 μl of streptomycin stock to it giving final working concentration of 100 μg ml^{-1}. Incubate the broth cultures at 28 °C in shaker for 2–3 days at 120 rpm. After sufficient growth of recipient strains, the *E. coli* strains are activated. Activate the *E. coli* strain containing donor (pHC60) plasmid in Luria broth + tetracycline (10 μg ml^{-1}). Similarly activate the *E. coli* strain with helper (pRK2013) plasmid in Luria broth + kanamycin (50 μg ml^{-1}). Incubate these *E. coli* strains at 37 °C overnight in shaker at 120 rpm.
2. After sufficient growth of all three types (recipient, donor and helper) of strains, triparental mating (conjugation) is performed. Take 1 ml broth of recipient strain in 1.5 ml microcentrifuge tube and centrifuge at 9000 × *g* for 1 min. Remove the supernatant; add 1 ml of 10 mM $MgSO_4$ to resuspend the pellet and vortex. Similarly, wash the pellet twice for removing antibiotics and resuspend the pellet finally in 100 μl of 10 mM $MgSO_4$. Similarly the helper and donor strains are processed

and suspension of 100 μl is prepared. Now add 45 μl of recipient strain, 15 μl of donor strain and 15 μl of helper strain on a TY plate and uniformly spread bacterial cells with the help of L-shaped spreader. Also streak the three types of strains individually on TY control plate and incubate all plates at 28 °C for overnight. The control plate is prepared to ensure growth of all three mating strains in conjugation plate.

3. After overnight incubation, for screening the trans-conjugants the TY plate is flooded with 3 ml of 10 mM $MgSO_4$ and the cells are scraped. One ml of conjugation mixture is transferred to 1.5 ml microcentrifuge tube (*see* **Note 8**). Serially dilute this homogenous cell suspension at different concentrations- 1:10, 1:100 and 1:1000. Spread 50 μl from each dilution on RDM selective medium containing antibiotics streptomycin (Sm^{100}) and tetracycline (Tc^{10}) (*see* **Notes 9** and **10**). Incubate the RDM plates at 28 °C and wait for the colonies to appear. Observe the bacterial colonies for expression of GFP under fluorescent microscope (Fig. 4c). When excited with blue light of 488 nm GFP emits bright green fluorescence (507–515 nm) in visible spectrum. The tagged rhizobial strains expressing GFP are inoculated on selective host legume as described in Subheading 3.6. After about 6–8 weeks from day of inoculation the plants are harvested from pots to record nodulation and to check expression of GFP within nodules (Fig. 4d) formed to confirm the occupancy of microsymbiont. The GFP tagged RNB strains can also be used to study the symbiotic interactions between legume host and rhizobia.

4 Notes

1. In case the rhizospheric soil is insufficient to fill the pots for trap experiments, the sterile river sand can be mixed with it to fill the pots for rhizobia trapping. Or a suspension of rhizospheric soil can be made in sterile water and used for inoculating the roots of germinated seedlings to be transplanted in sterile soil.
2. To ensure proper surface sterilization of nodules the sixth washing water as well as the intact nodules can be streaked on CR-YEMA plate to ensure that no surface contaminants are left. Rhizobia generally do not absorb red dye while the contaminants strongly absorb the dye. However, now it is widely accepted that rhizobia too can take red dye if cultured for longer period on media containing CR.
3. For preparing the cell templates bacterial cultures must be pure and freshly activated. For high exopolysaccharide producing strains wash the pellet two to three times with saline. Cell

templates must be used within 1 month for PCR reactions. Vortex the cell templates prior to adding to the PCR reaction mix to ensure resuspension of the cells.

4. If salts and unincorporated dyes are not removed from the sequencing reaction, they will compete with the extension fragments during electro kinetic injection resulting in weak signals.
5. The procedure for scarification varies with the type of seeds; for few crop legumes hot water treatment is sufficient, whereas for seeds of few wild legumes acid-treatment is required to break their dormancy. The duration of sterilization with a particular sterilizing agent depends upon the size and number of seeds being processed. The use of water agar or moistened filter paper for germination depends upon the better response of seeds while pretesting the viability of seeds.
6. Polyvinyl chloride tubes with lids are autoclaved and inserted in the middle of pots aseptically for adding water and nutrient solution periodically. Thoroughly washed, sterile marble/plastic beads are used to cover the soil surface to a depth of 15 mm in order to protect the soil from contaminates.
7. The biomass of treated/inoculated plants is compared with two types of controls (N+ and N−). More biomass of treated (nodulating) plants in comparison to N+ control indicates effective nodulation and nitrogen fixation by tested rhizobial strains. No nodulation should be observed in the N+ and N− control plants. Chances of contamination occur if seeds and soil mixture are not properly sterilized. To get authentic results of cross-infection tests the rhizobial strains can be tagged with *gfp* reporter genes and the expression of GFP within root nodules formed by infection of these tagged bacterial strains confirms their localization.
8. The homogenous triparental mating mixture can be stored at 4 °C and can be used to repeat serial dilution if required. In this experiment through triparental mating/conjugation the donor plasmid (pHC60) containing the *gfp* gene is transferred to the recipient RNB strain using the conjugation/DNA transfer machinery encoded by the helper plasmid (pRK2013).
9. For preparing solid agar plates with antibiotics the autoclaved media is cooled down and kept in water bath at 55–60 °C to prevent heat inactivation of antibiotics. When the temperature of the media is uniformly maintained at 55–60 °C, the antibiotics stocks are added as per dilution required. For example to prepare 500 ml RDM to screen trans-conjugants add 500 μl each of the streptomycin (100 mg ml^{-1}) and tetracycline (10 mg ml^{-1}) antibiotic stock solution this will give final

working concentration of 100 μg ml^{-1} and 10 μg ml^{-1} respectively.

10. The transformed cells of the recipient rhizobial strain will be able to grow on selective RDM containing antibiotics Sm^{100} and Tc^{10}. The donor and helper *E. coli* strains will not grow on the RDM as they are unable to metabolize the sucrose present in it. The rhizobial strains are generally resistant to streptomycin and due to insertion of the pHC60 plasmid the transformed rhizobial strains acquires resistance to tetracycline also. In case the recipient rhizobial strains to be tagged are sensitive to streptomycin any other antibiotic marker can be used.

Acknowledgments

The authors are thankful to the following funding agencies for financial assistance: University Grants Commission, New Delhi; DBT, Govt. of India; DST-SERB; and Crawford Fund Award-ATSE, Australia conferred to HSG and NT for advance training at CRS, Murdoch University, Perth, WA. We sincerely acknowledge Prof. Ann Hirsch, University of California, Los Angeles, USA and American Society for Microbiology for the ASM-IUSSTF Indo-US Research Professorship award to NT for GFP-related studies.

References

1. Lewis G, Schrire B, Mackinder B, Lock M (eds) (2005) Legumes of the world. Royal Botanic Gardens, Kew
2. Sprent JI (2009) Legume nodulation: a global perspective. Wiley-Blackwell, Oxford
3. Azani N, Babineau M, Bailey CD, The Legume Phylogeny Working Group (LPWG) et al (2017) A new subfamily classification of the Leguminosae based on a taxonomically comprehensive phylogeny. Taxon 66:44–77
4. Sprent JI, Ardley JK, James EK (2013) From north to south: a latitudinal look at legume nodulation processes. S Afr J Bot 89:31–41
5. Sprent JI, Ardley JK, James EK (2017) Biogeography of nodulated legumes and their nitrogen-fixing symbionts. New Phytol 215:40–56
6. Andrews M, Andrews ME (2017) Specificity in legume-rhizobia symbioses. Int J Mol Sci 18:705
7. Gyaneshwar P, Hirsch AM, Moulin L et al (2011) Legume-nodulating betaproteobacteria: diversity, host range, and future prospects. Mol Plant Microbe Interact 24:1276–1288
8. Peix A, Ramírez-Bahena MH, Velázquez E et al (2015) Bacterial associations with legumes. Crit Rev Plant Sci 34:17–42
9. Shamseldin A, Abdelkhalek A, Sadowsky MJ (2017) Recent changes to the classification of symbiotic, nitrogen-fixing, legume-associating bacteria: a review. Symbiosis 71:91–109
10. Herridge DF (2008) Inoculation technology for legumes. In: Dilworth MJ, James EK, Sprent JI, Newton WE (eds) Nitrogen-fixing leguminous symbioses. Springer, Dordrecht, pp 77–109
11. Baldwin IL, Fred EB (1929) Nomenclature of the root-nodule bacteria of the Leguminosae. J Bacteriol 17:141–150
12. Le Quéré A, Tak N, Gehlot HS et al (2017) Genomic characterization of *Ensifer aridi*, a proposed new species of nitrogen-fixing rhizobium recovered from Asian, African and American deserts. BMC Genomics 18:85
13. Mahato NK, Gupta V, Singh P et al (2017) Microbial taxonomy in the era of OMICS: application of DNA sequences, computational tools and techniques. Antonie Van Leeuwenhoek 110:1357–1371

14. Richardson AE, Viccars LA, Watson JM et al (1995) Differentiation of *Rhizobium* strains using the polymerase chain reaction with random and directed primers. Soil Biol Biochem 27:515–524
15. Stępkowski T, Moulin L, Krzyżańska A et al (2005) European origin of *Bradyrhizobium* populations infecting lupins and serradella in soils of Western Australia and South Africa. Appl Environ Microbiol 71:7041–7052
16. Martens M, Dawyndt P, Coopman R et al (2008) Advantages of multilocus sequence analysis for taxonomic studies: a case study using 10 housekeeping genes in the genus *Ensifer* (including former *Sinorhizobium*). Int J Syst Evol Microbiol 58:200–214
17. Gehlot HS, Panwar D, Tak N et al (2012) Nodulation of legumes from the Thar desert of India and molecular characterization of their rhizobia. Plant Soil 357:227–243
18. Gehlot HS, Tak N, Kaushik M et al (2013) An invasive *Mimosa* in India does not adopt the symbionts of its native relatives. Ann Bot 112:179–196
19. Tak N, Gehlot HS, Kaushik M et al (2013) Genome sequence of *Ensifer* sp. TW10; a *Tephrosia wallichii* (Biyani) microsymbiont native to the Indian Thar Desert. Stand Genomic Sci 9:304–314
20. Panwar D, Tak N, Gehlot HS (2014) Nodulated native legumes in an arid environment of Indian Thar Desert. In: Fulekar MH, Kale RK (eds) Recent trends in life sciences. IK International Publishing House Pvt. Ltd, India, pp 284–306
21. Sankhla IS, Meghwal RR, Tak N et al (2015) Phenotypic and molecular characterization of microsymbionts associated with *Crotalaria medicagenia*: a native legume of the Indian Thar desert. Plant Archives 15:1003–1010
22. Gehlot HS, Ardley J, Tak N et al (2016) High-quality permanent draft genome sequence of *Ensifer* sp. PC2, isolated from a nitrogen-fixing root nodule of the legume tree (Khejri) native to the Thar Desert of India. Stand Genomic Sci 11:43
23. Tak N, Awasthi E, Bissa G et al (2016) Multi locus sequence analysis and symbiotic characterization of novel *Ensifer* strains nodulating *Tephrosia* spp. in the Indian Thar Desert. Syst Appl Microbiol 39:534–545
24. Choudhary S, Meghwal RR, Sankhla IS et al (2017) Molecular characterization and phylogeny of novel diverse nitrogen fixing microsymbionts associated with *Vachellia* (*Acacia*) *leucophloea* in arid and semi arid regions of Rajasthan. Indian For 143:266–278
25. Sankhla IS, Tak N, Meghwal RR et al (2017) Molecular characterization of nitrogen fixing microsymbionts from root nodules of *Vachellia* (*Acacia*) *jacquemontii*, a native legume from the Thar Desert of India. Plant Soil 410:21–40
26. Rathi S, Gaur S, Tak N et al (2017) Genetically diverse root nodule bacteria associated with *Alysicarpus vaginalis* from alkaline soil of Rajasthan, India. Plant Archives 17:495–505
27. Choudhary S, Tak N, Gehlot HS (2018) Phylogeny and genetic diversity assessment of *Ensifer* strains nodulating *Senegalia* (*Acacia*) *senegal* (L.) Britton. In arid regions of Western Rajasthan, India. Microbiology 87:127–142
28. Sankhla IS, Meghwal RR, Choudhary S et al (2018) Molecular characterization of microsymbionts associated with root nodules of *Crotalaria burhia* Buch.-ham. Ex Benth., a native keystone legume species from Thar Desert of India. Indian J Exp Biol 56:373–384
29. Ojha A, Tak N, Rathi S et al (2017) Molecular characterization of novel *Bradyrhizobium* strains nodulating *Eriosema chinense* and *Flemingia vestita*, important unexplored native legumes of the sub-Himalayan region (Meghalaya) of India. Syst Appl Microbiol 40:334–344
30. Somasegaran P, Hoben HJ (1994) Handbook for rhizobia: methods in legume-rhizobium technology. Springer-Verlag, New York
31. Howieson JG, Dilworth MJ (eds) (2016) Working with rhizobia. Australian Centre for International Agricultural Research, Canberra
32. Morón B, Soria-Díaz ME, Ault J et al (2005) Low pH changes the profile of nodulation factors produced by *Rhizobium tropici* CIAT899. Chem Biol 12:1029–1040
33. Weisburg WG, Barns SM, Pelletier DA et al (1991) 16S ribosomal DNA amplification for phylogenetic study. J Bacteriol 173:697–703
34. Yanagi M, Yamasato K (1993) Phylogenetic analysis of the family Rhizobiaceae and related bacteria by sequencing of 16S rRNA gene using PCR and DNA sequencer. FEMS Microbiol Lett 107:115–120
35. Gaunt MW, Turner SL, Rigottier-Gois L et al (2001) Phylogenies of *atpD* and *recA* support the small subunit rRNA-based classification of rhizobia. Int J Syst Evol Microbiol 51:2037–2048
36. Turner SL, Young JPW (2000) The glutamine synthetases of rhizobia: phylogenetics and evolutionary implications. Mol Biol Evol 17:309–319
37. Stępkowski T, Czaplińska M, Miedzinska K et al (2003) The variable part of the *dna*K gene as an alternative marker for phylogenetic

studies of rhizobia and related alpha Proteobacteria. Syst Appl Microbiol 26:483–494

38. Chaintreuil C, Boivin C, Dreyfus B et al (2001) Characterization of the common nodulation genes of the photosynthetic *Bradyrhizobium* sp. ORS285 reveals the presence of a new insertion sequence upstream of *nodA*. FEMS Microbiol Lett 194:83–86
39. Haukka K, Lindström K, Young JPW (1998) Three phylogenetic groups of *nodA* and *nifH* genes in *Sinorhizobium* and *Mesorhizobium* isolates from leguminous trees growing in Africa and Latin America. Appl Environ Microbiol 64:419–426
40. Zhang XX, Turner SL, Guo XW et al (2000) The common nodulation genes of *Astragalus sinicus* rhizobia are conserved despite chromosomal diversity. Appl Environ Microbiol 66:2988–2995
41. Laguerre G, Nour SM, Macheret V et al (2001) Classification of rhizobia based on *nodC* and *nifH* gene analysis reveals a close phylogenetic relationship among *Phaseolus vulgaris* symbionts. Microbiology 147:981–993
42. Perret X, Broughton WJ (1998) Rapid identification of *Rhizobium* strains by targeted PCR fingerprinting. Plant Soil 204:21–34
43. Howieson JG, Loi A, Carr SJ (1995) *Biserrula pelecinus* L.-a legume pasture species with potential for acid, duplex soils which is nodulated by unique root-nodule bacteria. Aust J Agric Res 46:997–1009
44. Cheng HP, Walker GC (1998) Succinoglycan is required for initiation and elongation of infection threads during nodulation of alfalfa by *Rhizobium meliloti*. J Bacteriol 180:5183–5191
45. Figurski DH, Helinski DR (1979) Replication of an origin-containing derivative of plasmid RK2 dependent on a plasmid function provided in trans. Proc Natl Acad Sci U S A 76:1648–1652
46. Cappuccino JG, Sherman N (2007) Microbiology: a laboratory manual. Dorling Kindersley Pvt. Ltd, License of Pearson Education, New Delhi, India
47. Cheng HR, Jiang N (2006) Extremely rapid extraction of DNA from bacteria and yeasts. Biotechnol Lett 28:55–59
48. Kumar S, Stecher G, Tamura K (2016) MEGA7: molecular evolutionary genetics analysis version 7.0 for bigger datasets. Mol Biol Evol 33:1870–1874
49. Yates RJ, Howieson JG, Nandasena KG et al (2004) Root-nodule bacteria from indigenous legumes in the north-west of Western Australia and their interaction with exotic legumes. Soil Biol Biochem 36:1319–1329

Chapter 13

Detection and Quantification of Nitrifying Bacteria Using Real-Time PCR

Engil I. P. Pereira and Marcelo C. M. Teixeira Filho

Abstract

Nitrification is the microbial-mediated transformation of ammonium (NH_4^+) into nitrate (NO_3^-). Many plant species depend on the availability of NO_3^- as the main source of nitrogen (N). On the other hand, because NO_3^- is highly mobile in the soil profile, its excess concentration can cause environmental pollution. Nitrification can be estimated at the process level, but with the development of molecular techniques it is also possible to estimate the abundance of nitrifying bacteria in the soil. Hence, in this chapter we describe the procedure for detection and quantification of nitrifying bacteria in soil samples using real-time quantitative polymerase chain reaction (PCR).

Key words Nitrification, *Nitrosomonas europaea*, Nitrifying gene, DNA extraction, Amplification efficiency

1 Introduction

Nitrogen (N) is the mineral element required in largest quantity by plants; it plays a central role in plant metabolism as a constituent of proteins, nucleic acids, chlorophyll, coenzymes, phytohormones, and secondary metabolites [1, 2]. In order to achieve efficient growth, development, and reproduction, plants require adequate, but not excessive, amounts of N [1, 2]. Therefore, low soil N availability or a decline in root uptake capacity will negatively affect plant productivity and ecological competiveness [1].

In agricultural crops, N is the most limiting nutrient and accounts for a significant share in the production cost of these crops. Noncrop plants rely mostly on microbial-mediated processes such as mineralization and nitrification, for N supply. However, the efficiency of N use by plants is still low due to various loss factors such as ammonia volatilization, NO_3^- leaching, erosion, nitrification, denitrification, and nitrous oxide emissions [3].

The reduction of nitrification rates in the soil can contribute to improve nitrogen use efficiency (NUE) due to the reduction of

Kapuganti Jagadis Gupta (ed.), *Nitrogen Metabolism in Plants: Methods and Protocols*, Methods in Molecular Biology, vol. 2057, https://doi.org/10.1007/978-1-4939-9790-9_13,

nitrate (NO_3^-) leaching, reduction of nitrous oxide (N_2O) emission and a higher supply of N in the ammoniacal form to the plant. Nitrate is soluble in water and has low binding energy with soil colloids [4], which contributes to the contamination of surface and subsurface waters. Another aspect of NO_3^- leaching is the favoring of cation leaching. The anions SO_4^{2-} and NO_3^- from the mineralization of organic residues and fertilizers may promote, under certain conditions, the downward movement of the basic cations (Ca^{2+}, Mg^{2+} and K^+) in the profile, due to the ionic association in the soil solution (formation of ion pairs) with the momentary charge neutralization [5, 6]. Thus, NO_3^- leaching could also contribute to increase cation leaching as shown in some studies [7–9].

High levels of NO_3^- in soil can also result in accumulation of this N form in plant tissue, which may compromise the quality of plants for human and animal consumption [10]. Moreover, under O_2 deficiency conditions, some bacteria use N-NO_3^- as the final electron acceptor in the respiratory chain, reducing it to N_2 in the denitrification process [11]. One of the intermediate gases of this process is nitrous oxide (N_2O), which can affect the ozone layer, besides contributing to global warming [12].

Ammonium oxidation is the key process of nitrification. Only a few known groups of microorganisms are able to carry it out [13]. Traditionally, beta and gamma-proteobacteria have been recognized as exclusive agents in this process [14]. Heterotrophic organisms such as fungi, chemotrophic bacteria, and even plants can also perform nitrification by different biochemical processes; however, chemoautotrophic nitrification is the main route of nitrification in soil [15, 16].

For nitrification to occur, optimal population conditions of nitrifying microorganisms, pH, temperature, oxygen, moisture, and substrate availability are important. The optimum temperature for nitrification seems to be between 25 and 35 °C, at which point nitrifying bacteria develop best [16].

Clay soils generally have higher fluxes and mineralization nitrification than sandy soils. Clay limits N mineralization and nitrification by reducing the diffusion of organic matter. During periods of high humidity, the clay protects the organic N from the attack of microorganisms. Cycles of moistening and drying, however, break the physical protection promoted by the clay. The presence of a larger contact surface increases the establishment of bacterial biomass by providing fixation sites and concentrating diluted nutrients in solution, which favors nitrification [16]. Therefore, in general, nitrification appears to be faster in clay soils.

High humidity with reduced soil aeration impairs nitrification, which depends on oxygen. The nitrification rate in the soil is maximum in the field capacity [16]. Evidence shows that high humidity is more harmful than low humidity. Depending on the soil texture, the same gravimetric water content may result in different potential potentials, which implies different water availability for soil microorganisms, affecting the activity of nitrifying bacteria in low water content, both by cell dehydration by the substrate limitation [16].

In turn, the quality of organic matter directly affects nitrification, especially its C:N ratio. High C:N ratio favors the immobilization of ammonium, resulting in a smaller amount of the ion available for nitrifying organisms [16].

It is also known that soil pH is the main regulator of nitrification. The process occurs at pH values between 5 and 10, with an optimum value around 8.5 [16]. However, even the most acidic soils (pH $\approx$ 3) have reported the occurrence of nitrification. The substrate for the ammonia monooxygenase enzyme is ammonia (NH_3) and not NH_4^+ [17]. Excessively acidic environments may limit the formation of NH_3. Some researches have shown an increase in nitrification intensity due to liming [5, 18].

The nitrification reaction releases hydrogen (H^+), which results in soil acidification [16]. Over time, this acidification may end up limiting the process of nitrification in the soil. Thus, correction of soil acidity through liming could stimulate nitrification, especially in no-tillage systems, where surface liming is used. In this case the soil surface area that also receives the nitrogen fertilizer would present ideal conditions for nitrification [5].

Nitrification occurs rapidly in most agricultural soils with adequate pH, with most of the NH_4^+ being nitrified in 4 weeks. As a cation, ammonium is bound electrostatically to negatively charged surfaces of clay minerals and functional groups of soil organic matter (mainly carboxylic groups), which limits leach losses [19, 16]. The same does not occur with NO_3^-, which is highly mobile in the soil, and therefore more subject to leach losses. Thus, maintaining nitrogen in the form of NH_4^+ extends its residence time in the root zone, remaining available to the plant for longer [19], thus allowing greater N absorption by plants.

It is noteworthy that in similar proportions in the cultivation medium or soil, the plant absorbs more NH_4^+, probably because the assimilation of N occurs without the energy expenditure of the assimilative NO_3^- reduction [20], thus being more interesting for plant due to the lower consumption of ATP. However, excessive N uptake as NH_4^+ will cause a reduction in the development of the plant, given that the plant would not have the capacity to incorporate all the N from the NH_4^+ in the C skeletons (for lack of skeletons) in a timely manner, and with alkaline (high pH) would have the conversion of NH_4^+ into NH_3, which would result in phytotoxicity, that is, the pH gradient would be dissipated in the plant [20]. However, the largest N uptake by plants occurs in the form of NO_3^-, due to the nitrification process that makes this form of mineral N more available in the soil.

Therefore, is important to detect and quantify the nitrification caused by bacteria in the soil, aiming at higher nitrogen use efficiency, with a reflection on N metabolism in plants.

Here, we describe a procedure to detect and quantify nitrifying bacteria in soil samples using real-time qPCR. The primers and PCR conditions were developed by Holmes et al. [21] and Okano et al. [22]. We complement their descriptions by describing the procedure for production of positive standard controls, handling of soil samples, and issues that may arise during the procedure.

2 Materials

2.1 Soil Sampling

We suggest referring to Pennock et al. [23] for appropriate determination of sampling design. Materials usually used for soil samples include the following:

1. Soil auger.
2. Bucket for soil mixing.
3. Spatula.
4. Ethanol.
5. Disposable gloves.
6. Task wipers.
7. Store soil samples in DNA-free bags or containers.
8. Coolers.
9. Ice packs.
10. Soil moisture sensor.

2.2 DNA Extraction

Several commercial DNA extraction kits are available for soil samples. Examples:

1. MP Bio—FastDNA™ SPIN Kit for Soil (Catalog number: 16560200).
2. Norgen—Soil DNA Isolation Kit (Catalog number: 26500).
3. Qiagen—DNeasy PowerLyzer PowerSoil (Catalog number: 12855–50).
4. Zymo Research—Quick-DNA Fecal/Soil Microbe Kits (Catalog number: D6012).

The performance of each kit may vary with soil type and microbial biomass. Ideally, the user should screen the available kits for the best concentration and quality of DNA. Furthermore, it is important to use the same DNA extraction procedure within the same study to avoid variation from kit-specific extraction efficiency.

For quantification of DNA concentration, two instruments are commonly used:

1. NanoDrop™ (Thermo Scientific): This instrument measures DNA concentration through absorbance readings performed at 260 nm (A260) that is where DNA absorbs light most strongly.
2. Qubit Fluorometer (Invitrogen): This instrument requires the use of DNA-binding dye that allows for specific measurement of DNA.

2.3 Positive Control Standards

To calculate the abundance of nitrifying gene in a given sample, a standard curve using serial dilutions of positive control samples need to be included in the real-time PCR run. Materials needed are as follows:

1. *Nitrosomonas europaea* (ATCC 19718) containing *amo*A gene (GenBank: Z97833).
2. DNA extraction kit: QIAprep Spin Miniprep Kit (Catalog number: 27104).
3. Forward primer (A189): GGHGACTGGGAYTTCTGG designed by Holmes et al. [21].
4. Reverse primer (amoA-2R): CCTCKGSAAAGCCTTCTTC designed by Okano et al. [22].
5. Conventional PCR system.
6. GoTaq® DNA Polymerase from Promega (Catalog number M3001).
7. DNase-free water.
8. Phage vector (e.g., M13 bacteriophage vectors).
9. T4 DNA Ligase from Thermo Scientific (Catalog number EL0014).
10. Competent *E. coli* cells from Millipore Sigma (Catalog number CMC0007).

2.4 Real-Time PCR

Instruments:

1. Real-Time qPCR system: There are several Real-Time PCR systems produced by different manufacturers. For this protocol, we will suggest the use of PCR Applied Biosystems® 7500 Real-Time PCR and matching reagents.
2. Centrifuge with plate adaptors.

Reaction components:

1. Power SYBR Green RT-PCR Master Mix by Applied Biosystems (Catalog number 4368711) contains the polymerase enzyme, nucleotides, pH buffers, and the SYBR Green dye that binds to double-stranded (ds) DNA, providing a

fluorescent signal that reflects the amount of dsDNA product generated during PCR.

2. Forward primer (A189): GGHGACTGGGAYTTCTGG designed by Holmes et al. [21].
3. Reverse primer (amoA-2R): CCTCKGSAAAGCCTTCTTC designed by Okano et al. [22].
4. DNA template.
5. Positive control standards.

2.5 Materials

1. 96-well plate for qPCR: MicroAmp™ Fast Optical 96-Well Reaction Plate with Barcode, 0.1 ml (Catalog number 4366932).
2. MicroAmp™ Optical Adhesive Film (Catalog number 4311971).
3. 96-well plate rack.
4. 2 ml centrifuge tube.
5. Pipettes (10, 100, 200, and 1000 μl).
6. Pipette tips.
7. Nuclease-free water.
8. Ice bucket.
9. Ice.
10. PCR workstation with UV light.

3 Methods

3.1 Soil Sampling

In order to avoid degradation of bacterial DNA and sample contamination, make sure to:

1. Maintain samples stored in a cooler with ice while sampling and transporting.
2. Clean tools and devices with ethanol or DNase-free cleaning products to avoid cross-sample contamination.
3. Extract the DNA immediately or store samples in −20 °C freezer for short-term storage.

The content of soil moisture per sample need to be assessed in order to correct and express the copies of nitrifying genes per mass of dry soil. To assess soil moisture, the researcher can collect this information in the field using portable devices or estimate gravimetric soil moisture in the laboratory using a subsample.

3.2 DNA Extraction

Carry out DNA isolation according to instructions provided by the kit's manufacturer. A minimum of 1 μg of DNA per gram of dry soil should be obtained per sample. If higher DNA concentrations need to be achieved, the user may consider increasing the initial sample amount or lysing period. Highly concentrated samples can be diluted with purified water before further application.

3.3 Preparation of Positive Control

1. Extract the genomic DNA from *Nitrosomonas europaea* (ATCC 19718) pure culture.
2. Use the extracted DNA as template in a regular PCR reaction to amplify copies of *amoA*.
3. Clone the PCR product into a vector (i.e., M13) by incubating them with T4 DNA Ligase.
4. Then the plasmid can be inserted into competent *E. coli* cells, by integrating the plasmids as they grow.
5. Stock cultures of *E. coli* containing the plasmid can be frozen, and thawed and regrown later.
6. Before performing real-time qPCR, the plasmid DNA can be extracted from the *E. coli* culture.
7. The concentration of plasmid DNA needs to be quantified using the available methods in the laboratory (e.g., NanoDrop™ or Qubit).
8. The number of *amoA* copies is calculated from the respective concentrations of extracted plasmid DNA.
9. Serial dilutions of stock plasmid DNA will be used to generate the standard curves during the real-time qPCR.

3.4 Real-Time PCR

1. About 30 min before start preparing the reactions:
   (a) Bring all *materials* to the workstation PCR hood and maintain them under UV light for 30 min.
   (b) Thaw samples on ice.
   (c) Thaw primers on ice.
   (d) Thaw standards on ice.
   (e) Calculate the volume of all solutions composing the master mix:

Component	Vol. per reaction (μl)
SYBR Green	10
Forward primer (A189; 5 μM)	1.2
Reverse primer (amoA-2R; 5 μM)	3.6
Water	0.8
Total per reaction	**15**

(f) Create a run-file on the computer using the qPCR software identifying all samples, standards, and non-template controls (*see* **Note 1**).

(g) On the run-file insert the PCR program: 15 s at 95 °C, and then 40 cycles consisting of 15 s at 95 °C, 30 s at 55 °C, and 31 s at 72 °C (set as data acquisition stage), followed by a dissociation stage of 15 s at 95 °C, 30 s at 60 °C, and 15 s at 95 °C [23].

(h) Save with date, gene, initials, and format ".sds".

2. Prepare a "master mix" based on calculations (e) (*see* **Note 2**).
3. Distribute total volume of master mix on all wells of the 96-well plate.
4. Quickly spin standards.
5. Add standard to designed wells (5 μl).
6. Spin samples.
7. Add samples to designed wells (5 μl).
8. Cover plate with adhesive film.
9. Spin for 60 s at 2000 × *g*.
10. Insert plate in the machine, upload running file, make sure all cells were correctly identified on the system's software, and the run is ready to be started.
11. At the end of the run, export your file as an Excel file (.xls).

4 Notes

1. When preparing the table (Subheading 3.4) make sure to include wells for standards, blanks, and nontemplate control (NTC) in addition to your samples, all in triplicate.
2. When preparing the master mix, calculate the volume of components considering an extra 5% of samples for potential pipetting errors.
3. The amplification efficiency (E) will be estimated by the instrument by using the slope of the standard curve according to the following formula: $E = (10^{-1/\text{slope}}) - 1$.
4. Acceptable r^2 or coefficient of determination of the standard curve should be >0.99.
5. The qPCR system will provide a cycle quantification value per sample, often called the "threshold cycle (Ct)." The Ct is a relative measure of the concentration of target in the PCR reaction and based on the Ct values of the positive controls, it will be possible to estimate the amount of *amoA* gene copies in each sample.

References

1. Hawkesford M, Horst W, Kichey T, Lambers H, Schjoerring J, Møller IS, White P (2012) Functions of macronutrients. Chapter 6. Marschner P. Marschner's mineral nutrition of higher plants, 3rd edn. Elsevier, London
2. Malavolta E, Vitti GC, Oliveira SA (1997) Evaluation of the nutritional status of plants: principles and applications, 2nd edn. Potafos, Piracicaba
3. Weiske A, Benckiser G, Herbert T, Ottow J (2001) Influence of the nitrification inhibitor 3,4-Dimethylpyrazole phosphate (DMPP) in comparison to dicyandiamide (DCD) on nitrous oxide emissions, carbon dioxide fluxes and methane oxidation during 3 years of repeated application in field experiments. Biol Fertil Soils 34:109–117
4. Whitehead DC (1995) Grassland nitrogen. CAB International, Wallingford
5. Foloni JSS, Rosolem CA, Garcia RA (2006) Effect of liming and ammonium in cotton. II. Concentration of cations and anions in soil solution and absorption of nutrients by plants. Rev Bras Ciênc Solo 30:433–442
6. Cahn MD, Bouldin DR, Cravo MS, Bowen WT (1993) Cation and nitrate leaching in an Oxisol of the Brazilian Amazon. Agron J 85:334–340
7. Di HJ, Cameron KC (2005) Reducing environmetal impacts of agriculture by using a fine particle suspension nitrification inhibitor to decrease nitrate leaching from grazed pastures. Agric Ecosyst Environ 109:202–212
8. Di HJ, Cameron KC (2004) Treating grazed pasture soil with a nitrification inhibitor, eco-nTM, to decrease nitrate leaching in a deep sandy soil under spray irrigation: a lysimeter study. N Z J Agric Res 47:351–361
9. Wu S, Wu L, Shi Q, Wang Z, Chen X, Li Y (2006) Effects of a new nitrification inhibitor 3,4-dimethylpyrazole phosphate (DMPP) on nitrate and potassium leaching in two soils. J Environ Sci 19:841–847
10. L'hirondel J, L'hirondel JL (2002) Nitrate and man: toxic, harmless or beneficial. CAB International, Wallingford
11. Dendooven L, Bonhomme E, Merckx R, Vlassak K (1998) Injection of pig slurry and its effects on the dynamics of nitrogen and carbon in a loamy soil under laboratory conditions. Biol Fertil Soils 27:5–8
12. Rochette P, Angers DA, Chantigny MH, Bertrand N, Côté D (2004) Carbon dioxide and nitrous oxide emissions following fall and spring applications of pig slurry to agricultural soil. Soil Sci Soc Am J 68:1410–1420
13. Kleineidam K, Košmrlj K, Kublik S, Palmer I, Pfab H, Ruser R, Fiedler S, Schloter M (2011) Influence of the nitrification inhibitor 3,4-dimethylpyrazole phosphate (DMPP) on ammonia-oxidizing bacteria and archaea in rhizosphere and bulk soil. Chemosphere 84:182–186
14. Bock E, Wagner M (2006) Oxidation of inorganic nitrogen compounds as an energy source. In: Dworkin M (ed) The prokaryotes. Springer, New York
15. Moreira FMS, Siqueira JO (2006) Microbiology and soil biochemistry, 2nd edn. Federal University of Lavras, Lavras
16. Sarahwat KL (2008) Factor affecting nitrification in soils. Commun Soils Sci Plant Anal 39:1436–1446
17. De Boer W, Kowalchuk GA (2001) Nitrification in acid soils: micro-organisms and mechanisms. Soil Biol Biochem 33:853–966
18. Silva CA, Vale FR (2000) Availability of nitrate in Brazilian soils under the effect of liming and sources and doses of nitrogen. Pesq Agrop Brasileira 35:2461–2471
19. Subbarao GV, Ito O, Sahrawat KL, Berry WL, Nakahara K, Ishikawa T, Watanabe T, Suenaga K, Rondon M, Rao IM (2006) Scope and strategies for regulation of nitrification in agricultural systems: challenges and opportunities. Crit Rev Plant Sci 25:302–335
20. Prado RM (2008) Plant Nutrition. UNESP, São Paulo
21. Holmes AJ, Costello A, Lidstrom ME, Murrell JC (1995) Evidence that particulate methane monooxygenase and ammonia monooxygenase may be evolutionarily related. FEMS Microbiol Lett 132:203–208
22. Okano Y, Hristova KR, Leutenegger CM, Jackson LE, Denison RF, Gebreyesus B, Scow KM (2004) Application of real-time PCR to study effects of ammonium on population size of ammonia-oxidizing bacteria in soil. Appl Environ Microbiol 70:1008–1016
23. Pennock D, Yates T, Braidek J (2007) Soil sampling designs. Soil sampling and methods of analysis. CRC Press, Canadian Society of Soil Science, Boca Raton, FL

Chapter 14

Extracellular Matrix Proteome: Isolation of ECM Proteins for Proteomics Studies

Eman Elagamey, Kanika Narula, Niranjan Chakraborty, and Subhra Chakraborty

Abstract

Understanding molecular mechanisms and cellular metabolism in varied plant processes necessitates knowledge of the expressed proteins and their subcellular distribution. Spatial partitioning of organelles generates an enclosed milieu for physiochemical reactions designed and tightly linked to a specific organelle function. Of which, extracellular matrix (ECM)/cell wall (CW) is a dynamic and chemically active compartment. The ECM proteins are organized into complex structural and functional networks involved in several metabolic processes, including carbon and nitrogen metabolism. Organellar proteomics aim for comprehensive identification of resident proteins that rely on the isolation of highly purified organelle free from contamination by other intracellular components. Extraction and isolation of plant ECM proteins features key caveats due to the lack of adjoining membrane, the presence of a polysaccharide–protein network that traps contaminants, and the existence of high phenolic content. Furthermore, due to diverse biochemical forces, including labile, weakly bound and strongly bound protein in the protein–polysaccharide matrix different elution procedures are required to enrich ECM proteins. Here, we describe a method that allows efficient fractionation of plant ECM, extraction of ECM proteins and protein profiling from variety of crop plants, including rice, chickpea and potato. This method can easily be adapted to other plant species for varied experimental conditions.

Key words Plant, Organellar proteomics, Extracellular matrix, Cell wall, Protein enrichment, Gel electrophoresis, Mass spectrometry

1 Introduction

Proteins epitomize informational macromolecules with diverse conformations, modifications and subcellular localization involved in variety of cellular reactions, cascades, and pathways with specific function. Spatial proteomics aims to study organelle specific protein and proteoforms that yield relatively simpler datasets from which biologically relevant information can be extracted as compared to whole-cell proteomes [1]. Nitrogen is one of the most important nutrients for plant growth and a major constituent of proteins that regulate various biological processes [2]. Extracellular matrix (ECM)

Kapuganti Jagadis Gupta (ed.), *Nitrogen Metabolism in Plants: Methods and Protocols*, Methods in Molecular Biology, vol. 2057, https://doi.org/10.1007/978-1-4939-9790-9_14,

is a reservoir of organic carbon and a major nitrogen sink having nitrogen content ranging from 0.4 to 2.2% in primary wall of several plant species [3, 4] It is highly organized supramolecular assembly of complex structural components unique to the specific tissue, organ and taxonomic clades. It also features meshwork of cross-linked proteins associated with biophysical and biochemical cues involved in cell communication, signaling, proliferation, survival and differentiation [5]. The intricate balance among carbon and nitrogen metabolism, ECM biogenesis and ECM protein turnover play a critical role in plant development and response to extra and intracellular stimuli [6]. However, the physiochemical properties of ECM proteins: large size, basic in nature, heavily glycosylated, highly cross-linked, poor solubility rendered its biochemical analyses a challenging task. ECM turnover, fractionation, quality and purity are of the utmost importance for its successful analysis. Methods of ensuring ECM purity include assays for marker enzyme activities. Its composition reflects the sublayering feature of primary and secondary wall that eases monitoring of the isolation procedure by transmission electron microscopy [7]. Plant ECM proteomics has considerably developed since the first plant ECM proteome was published [8]. Several studies of plant ECM protein complements have added to our understanding in a wide range of species [5, 9, 10].

Perhaps the primitive method of plant ECM protein isolation involves the direct extraction in solution containing $CaCl_2$ gradients and plasmolysis of plasma membrane to shrink ECM [11, 12]. Major drawbacks of this non-disruptive method are low yield of ECM proteins and the possibility of contamination with cytosolic and membrane proteins [13]. Shortcomings of above-mentioned method highlight the need for modification, optimization, and refinement of isolation protocols utilizing destructive methods which may suit different tissues/species. Another isolation protocol, include disruptive method in which cells were passed through N_2 disruption bomb or were homogenized in cell disrupter followed by sedimentation ECM fraction for further analysis [14–16]. ECM proteins were categorized based on their physiochemical association with wall components and extraction methods into water soluble and loosely ionically bound, tightly ionically bound, and covalently bound [17]. Ionically bound ECM proteins were eluted by incubating cultured cells in 1 M NaCl buffer followed by homogenization, vacuum infiltration, and extraction. 0.2 M $CaCl_2$ and potassium phosphates were used to extract tightly ionically bound (fraction 2) ECM proteins. However, covalently bound ECM proteins were eluted by treating wall fraction with mixture of cellulase/macerozyme to digest wall carbohydrates [18].

Owing to the fact that ECM is rich in polyphenolics and ECM proteins are mostly basic glycoproteins, their separation is best achieved by one-dimensional electrophoresis (1-DE) and two-dimensional electrophoresis (2-DE) [19]. Gels can then be

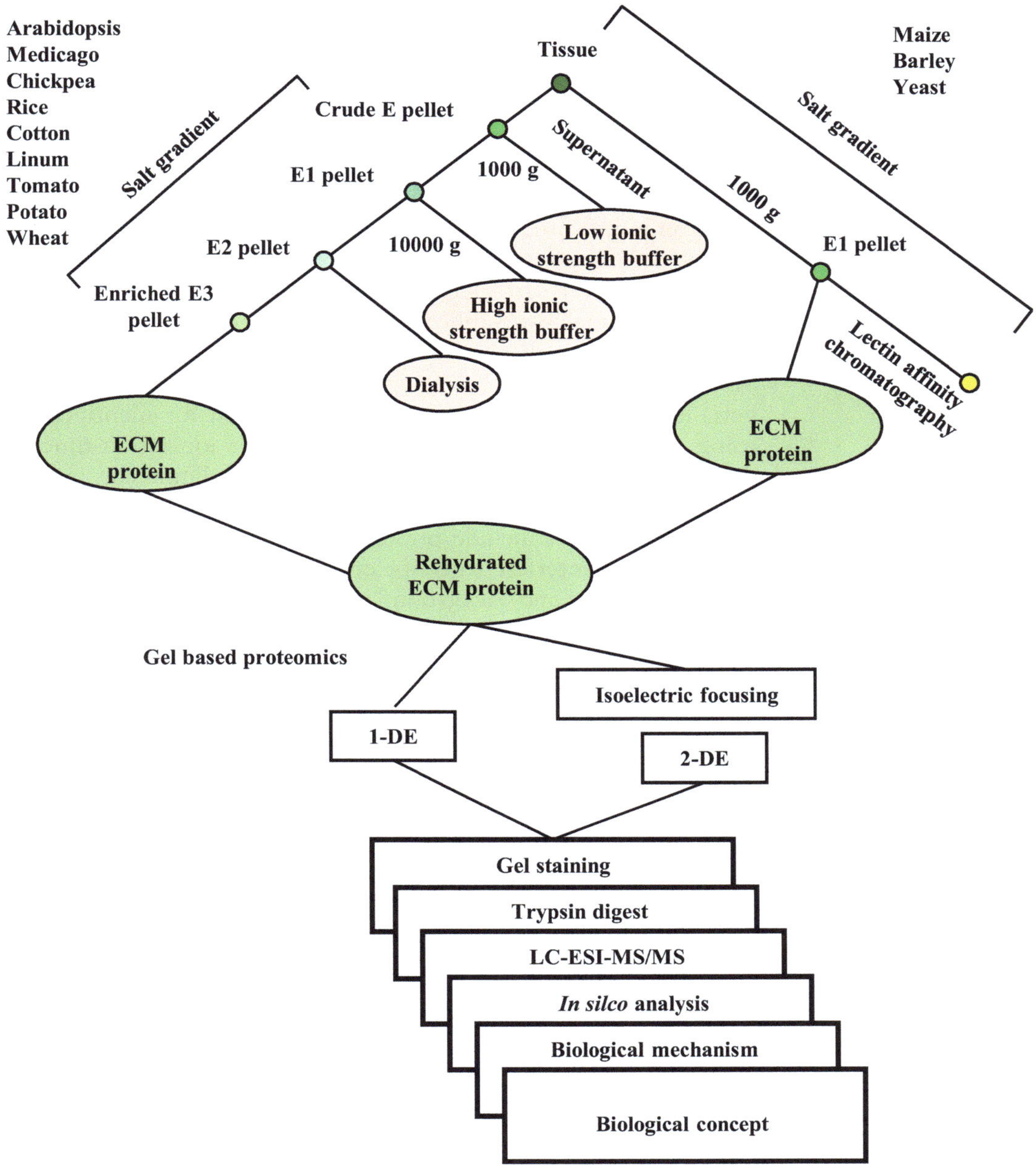

Fig. 1 Schematic representative of ECM isolation, protein extraction and proteomics analysis

stained using different reagents with varied sensitivities followed by excision and enzymatic digestion of protein bands to generate a defined set of peptides prior to mass spectrometry [20]. The method described here was established as a large-scale approach for comparative ECM proteomics of diverse plant species. It is based on our understanding of past 10 years ([5, 7, 21–26], *see* Subheading 4) and information found at http://cellwall.genomics.purdue.edu. Briefly, each step described in Fig. 1 can be optimized to yield reproducible results. The protocol for ECM isolation, protein extraction, and mass spectrometry coupled gel-based analysis described in this chapter is simple, reproducible, robust, and easy to carry out, that yield intact and pure ECM fraction and ECM proteins. The major steps involved mechanical disruption, density-based gradient centrifugation and series of stringent washes to remove contaminating proteins and extraction of ECM proteins with commonly used wall protein extractants, $CaCl_2$ (Subheadings 3.1 and 3.3). Rigorous purity assessment is performed at organellar and protein levels. Isolated ECM fractions are subjected to light microscopic analysis (Subheading 3.2) while extracted ECM protein enrichment is checked by enzyme assays (Subheading 3.4) and SDS–polyacrylamide gel electrophoresis (PAGE) (Subheading 3.5) gels. Enriched ECM proteins are further used for identification using either matrix-assisted laser desorption ionization-time of flight mass spectrometry (MALDI-TOF-MS) or liquid chromatography tandem mass spectrometry (LC-MS/MS) (Subheading 3.6). The major caveats include presence of polysaccharides and polyphenolics that interfere with isoelectric focusing in first dimension causing streaking and background staining in second dimension (*see* **Notes 19** and **20**). These difficulties were minimized resulting in ECM protein preparations free of contamination that yielded several hundred proteins.

2 Materials

2.1 Plant Material

1. Chickpea (*Cicer arietinum* L. cv. WR-315) seedlings are grown in an environmentally controlled growth room maintained at 25 ± 2 °C, 50 ± 5% relative humidity under 16 h photoperiod (270 μmol m^{-2} s^{-1} light intensity).
2. Three-weeks-old seedlings are used as experimental material and store at −80 °C after quick freezing in liquid nitrogen until further use or process immediately (*see* **Notes 1**and **2**).

2.2 Chemicals and Equipment

1. Ultrapure water (double-distilled, deionized) was used to prepare all reagents.
2. High grade reagents were used for buffer preparation and electrophoresis.
3. Mortar and pestle and Liquid N_2: to disrupt plant material mechanically (shoot and root) (*see* Subheading 3.1 and **Note 3**).
4. Spatula, funnels, flasks (*see* **Note 4**).
5. 2-D Quant kit (GE Healthcare).

6. Immobilized pH gradient (IPG; linear and nonlinear pH gradient from 3 to 10, 24 cm; 4 to 7, 13 cm and 24 cm length (GE Healthcare).
7. IPGphor Strip holders.
8. IPGphor apparatus for isoelectrofocusing of proteins Ettan™ IPGphore™ Isoelectric Focusing System.
9. Sodium dodecyl sulfate polyacrylamide gel electrophoresis (SDS-PAGE) molecular weight standards low range.
10. SDS-PAGE electrophoresis.
11. Glass plates for casting gels.
12. Silver stain Plus kit is used for staining protein spots.
13. Light microscope Nikon, Ninato, Tokyo, Japan.

2.3 Stock Solutions

1. 10 mM PMSF (phenylmethanesulfonyl fluoride) stock solution: dissolve 17.4 mg PMSF and add isopropanol to 10 mL. Aliquots are stored at −20 °C (*see* **Note 5**).
2. Store acetone at −20 °C.

2.4 Buffers

2.4.1 ECM Isolation

All buffers should be freshly prepared, either directly or as stock solutions, without adding protease inhibitors (it should be added fresh before use).

1. Homogenizing buffer: 5 mM K_3PO_4, pH 6.0, 5 mM DTT, 1 mM PMSF.

2.4.2 ECM Protein Extraction and Quantification

1. Protein extraction buffer: 200 mM $CaCl_2$, 5 mM DTT, 1 mM PMSF, 0.3% (w/w) PVPP.
2. Protein dilution buffer: 100 mM Tris–Cl (pH 8.5), 20% (v/v) glycerol, 4% (w/v) SDS, 20 mM DTT, and 1 mM PMSF.
3. 2-D rehydration buffer: 8 M urea, 2 M thiourea, 4% (w/v) CHAPS, 20 mM DTT, 0.5% (v/v) Pharmalyte, and 0.05% (w/v) bromophenol room temperature (25 °C). Aliquots are stored at −20 °C.

2.4.3 Enzyme Assays

Stock solutions for the buffers can be made a day before and stored at 4 °C. On the day of the experiment, the buffers can be diluted as required and used.

1. 0.2 M phosphate buffer pH 7.5: dissolve 1.4 g KH_2PO_4 in 50 mL water. Dissolve 17.4 g K_2HPO_4 in 500 mL water. Mix the two solutions, which give 550 mL of a 0.2 M phosphate buffer, pH 7.5. Store at −20 °C for several months.
2. 3% H_2O_2.

3. 0.1 mM orthovanadate: Add 1.839 g sodium orthovanadate to a falcon tube. Make up to 8 mL with distilled water. Adjust pH -10. If solution is yellow, place in boiling water until clear. Recheck pH and repeat as necessary. Store at −20 °C.

2.4.4 Light Microscopy

1. 0.1 M phosphate buffer pH 7.4: dissolve 35.5 g of sodium phosphate dibasic in a final volume of 500 mL of H_2O. Some crystallization will occur if stored at 4 °C. Stir the solution until crystals dissolve. Prepare sodium phosphate monobasic stock (0.5 M) by dissolving 30 g of anhydrous sodium phosphate monobasic in a final volume of 500 mL of H_2O. Take 80 mL of sodium phosphate dibasic stock (0.5 M) and add H_2O to give a final volume of 400 mL. Prepare 0.1 M sodium phosphate monobasic solution by dissolving 30 mL of sodium phosphate monobasic stock (0.5 M) to final volume of 150 mL in water. Bring the 0.1 M sodium phosphate dibasic solution to pH 7.4 by adding as much as needed of the 0.1 M sodium phosphate monobasic solution from.
2. 4% formaldehyde: dissolve paraformaldehyde powder into 0.1 M phosphate buffer pH 7.4 using a stirring hot plate. Once the solution is dissolved, it is quickly chilled on ice back to room temperature prior to use.

2.4.5 SDS–Polyacrylamide Gel

Two main gel-based methods (1-DE and 2-DE) are used to separate the ECM proteins using standard protocol for buffer preparations (*see* **Notes 6** and 7).

1. Gel casting buffers.

   For discontinuous gels, the buffers required to prepare resolving gel and stacking gel are different.

   (a) Resolving gel buffer: 1.5 M Tris–HCl, pH 8.8: dissolve 18.15 g of Tris base in 80 mL distilled water. Adjust pH 8.8 using 6 N HCl. Make final volume to 100 mL with distilled water.

   (b) Stacking gel buffer: 0.5 M Tris–HCl, pH 6.8: dissolve 6 g of Tris base in 80 mL distilled water. Adjust pH 6.8 using 6 N HCl. Make final volume to 100 mL with distilled water.
2. 1× Laemmli loading buffer: dissolve bromophenol blue (0.004%), 2-mercaptoethanol (10%), glycerol (20%), SDS (4%), and Tris–HCl 0.125 M. 2-mercaptoethanol should be added just before use.
3. Acrylamide–bis-acrylamide solution: Dissolve 30% acrylamide and 0.8% methylene bis-acrylamide (*see* **Note 8**) in 60 mL ddH_2O. Add a spatula of AG501-X8 (D) mixed-resin beads (Bio-Rad) and mix for about 30 min. Make up to 100 mL with

ddH_2O. Store at 4 °C, in a bottle wrapped with aluminum foil (*see* **Note 9**).

4. 20% (w/v) SDS (sodium dodecyl sulfate).
5. 10% APS (ammonium persulfate). Keep at 4 °C and/or make fresh.
6. 10× running buffer (1 L): 30 g Tris, 144 g glycine, and 10 g SDS dissolved in ddH_2O and make up to 1 L (*see* **Note 10**).
7. Coomassie Blue stain: 50% methanol, 10% (v/v) acetic acid, 0.4% (w/v) Coomassie Blue dye, and 40% ddH_2O.
8. Destaining solution: 50% methanol, 10% (v/v) acetic acid and 40% ddH_2O.
9. IEF rehydration buffer: 8 M urea, 2 M thiourea, 4% (w/v) CHAPS, 20 mM DTT, 0.5% (v/v) pharmalyte (pH either 4–7; GE Healthcare), and 0.05% (w/v) bromophenol blue.
10. Equilibration buffer: 6 M urea, 75 mM Tris–HCl (pH 8.8), 30% (v/v) glycerol and 2% (w/v) SDS. Aliquot and store at −20 °C until use.
11. Agarose sealing solution: Add 0.5% agarose in 1× SDS running buffer with a pinch of bromophenol added as tracking dye. Heat in a microwave oven so that agarose dissolves completely. Prepare just before use.

3 Methods

3.1 ECM Isolation

Buffers, glassware, tubes, and equipment should be precooled to 4 °C or kept on ice. Centrifuge rotors should be cooled down to the same temperature.

1. Ground 10 g tissue into powder in a prechilled mortar and pestle using liquid nitrogen (*see* **Notes 1–3**). Add 0.3% (w/w) PVPP to the ground powder and mix thoroughly (*see* **Note 11**).
2. Use spatula to transfer the ground powder immediately into a 50 mL ice-cold centrifuge tube containing chilled homogenizing buffer (*see* **Notes 12** and **13**).
3. Gently stir the contents for 2–3 min at 4 °C for the complete lysing of the cell and to preserve the ionic bonds and to dilute the ionic strength of the ECM.
4. Vortex this suspension for complete homogenization.
5. Pellet the homogenate by differential centrifugation with a fixed-angle rotor (1000 × *g*) at 4 °C for 5 min (*see* **Note 14** and Precaution).

6. Separate the supernatant and gently wash the pellet with ten times with deionized water to remove contaminants (*see* **Note 15**).
7. Resulting ECM pellet can be store on ice for further purity experiments or directly use for protein extraction.

3.2 Light Microscopy

1. Place 50 μL aliquot of the purified ECM fraction in 0.1 M phosphate buffer on glass slide.
2. Perform fixation in 4% (v/v) paraformaldehyde for about 1 h. Allow the suspension to fix at 4 °C.
3. Wash the slides having ECM fractions two to three times in 0.1 M phosphate buffer pH 7.4.
4. Add one to two drops of toluidine blue and allow to incubate for 2 min.
5. Rinse the sample by adding water and apply coverslip.
6. For microscopic analysis examine the slides by stereo zoom microscopy (Nikon, Ninato, Tokyo, Japan) and record the images (Fig. 2a).

3.3 ECM Protein Extraction and Quantification

1. Resuspend the ECM pellet completely in three volumes of extraction buffer containing $CaCl_2$.
2. Dissolve the pellet by repeated pipetting of the suspension.
3. Shake the ECM pellet with extraction buffer on a shaking platform for 2 h at 4 °C to remove any further contaminants and to release ECM proteins (*see* **Note 16**).
4. Following centrifugation at10,000 × *g* for 10 min at 4 °C, the ECM proteins gets separated from the insoluble fraction (*see* **Note 17**).
5. Filter the suspensions through 0.45 μm filter.
6. Concentrate the filtrate using Centricon YM3 and then dialyze overnight against 1000 volumes of deionized water with one change.
7. Dilute ECM proteins in dilution buffer and boil for 5 min.
8. Allow the protein samples to cool to room temperature (25 °C).
9. Add 9 volumes of 100% acetone to each tube and mix by inversion. Complete precipitation of proteins can be achieved by incubation for overnight at −20 °C (*see* **Notes 18** and **19**).
10. Remove the supernatant and recover the protein pellet by centrifugation at 10,000 × *g* for 10 min at 4 °C.
11. Wash the protein pellets two times with 80% acetone to remove excess SDS (*see* **Note 20**).

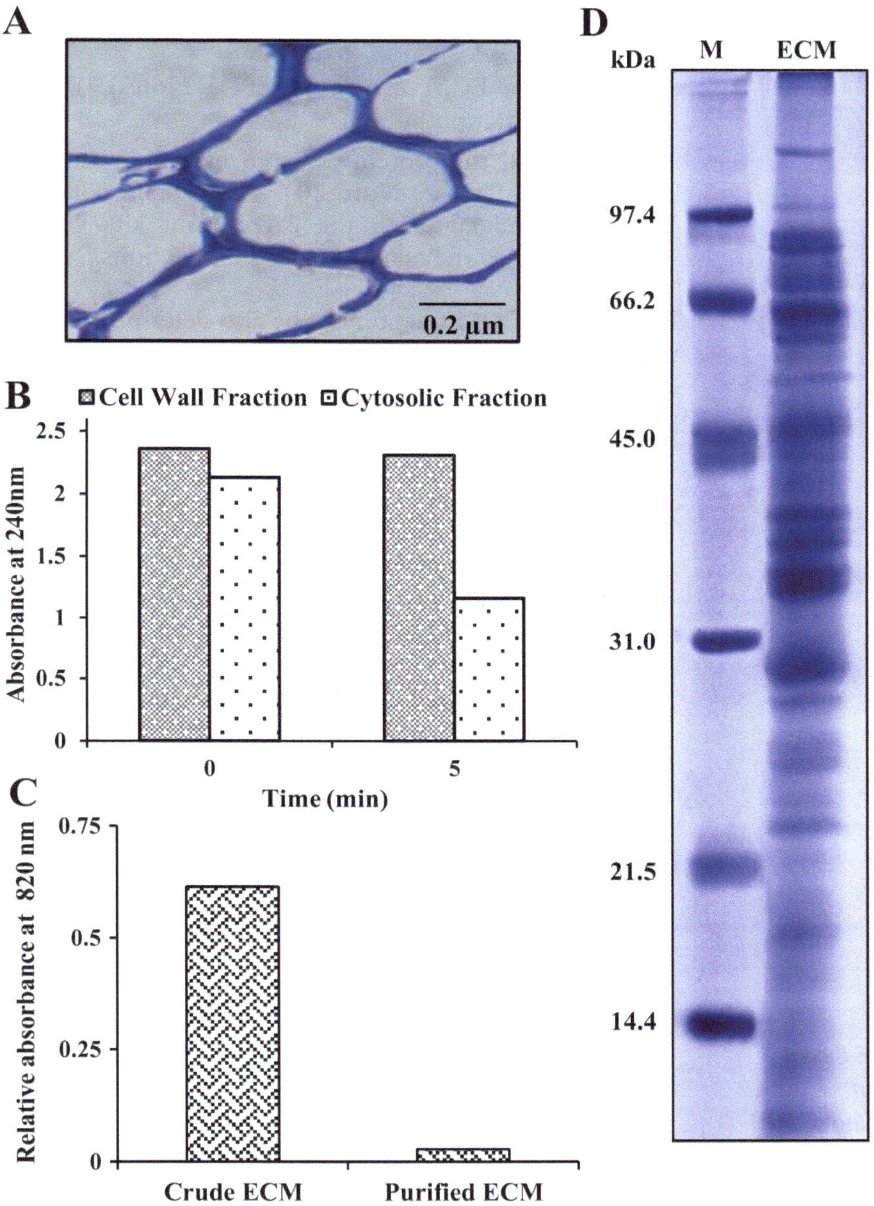

Fig. 2 Analysis of isolated chickpea ECM fraction and determination of its purity. (**a**) Stereo zoom micrograph of purified ECM fraction stained with toluidine blue. (**b**) Determination of catalase specific activity in ECM and cytosolic fraction (positive control). (**c**) Determination of vanadate inhibited H^+ ATPase activity. The plasma membrane fraction was used as positive control. (**d**) Representative analytical 1-D electrophoresis profile (12.5% SDS-PAGE, Coomassie Brilliant Blue) obtained from ECM protein extract. *M*, molecular mass protein standards; *ECM*, purified extracellular matrix protein

12. Centrifuge at 10,000 × *g* for 10 min at 4 °C Discard the supernatant.
13. Air-dry the ECM proteins pellet (*see* **Note 19**) and dissolve in 2-D rehydration buffer.
14. Resuspend the protein pellet in minimum amount of 2-DE sample buffer (*see* **Note 20**). The protein concentration can be determined using the 2-D Quant kit (GE Healthcare) according to the manufacturer's instructions.

3.4 Enzyme Assay

Purity of the ECM fraction can be also determined on the basis of marker enzyme activities specific for different organelles, viz., catalase for microbodies and vanadate inhibited H^+ ATPase for plasma membrane (*see* **Note 21** and Fig. 2b, c)

1. Determine the activity of catalase using 10 μg of organellar protein for each reaction. Prepare the reaction mixture by adding 50 μL of protein extract to 940 μL of 70 mM potassium phosphate buffer. Initiate the reaction by adding 10 μL of H_2O_2 (3% v/v) and monitor the decrease in absorbance at 240 nm for 5 min. Baseline correction can be done by subtracting the absorbance taken without the addition of H_2O_2.
2. Vanadate inhibited H+ ATPase activity can be determined with 20 μg of protein in 120 μL of assay buffer. Perform the assay in the presence and absence of freshly prepared 0.1 mM orthovanadate and boil the buffer prior to addition of 0.05% Triton X-100 (*see* **Note 22**). Initiate the reaction by addition of ATP and incubate at 37 °C for 30 min. Run blank lacking $MgSO_4$ in parallel and determine the release of inorganic phosphate. Measure the relative activity of the enzyme by taking the difference between the absorbance at 700 nm in absence and presence of orthovanadate.

3.5 1-DE Based Separation and Staining of ECM Protein

1. Clean the glass plates and spacers of the gel casting unit with detergent and water followed by ethanol. Assemble the gel cassette (16 × 18 cm; plates and spacers) on the stable even surface and attach this assembly in the gel electrophoresis apparatus (*see* **Note 23** and Fig. 2d).
2. Prepare polyacrylamide gel according to standard protocol (for 12.5% gel, resolving gel solution (30 mL): 12.5 mL acrylamide/bis-acrylamide solution, 7.8 mL resolving gel buffer, 150 μL 20% SDS, 200 μL 10% APS, 40 μL TEMED, and 9.6 mL ddH_2O). Add TEMED and APS at the end.
3. Pour the gel solution in the plates assembled with spacers. To maintain an even and horizontal resolving gel surface, layer the surface with water or isopropanol or butanol.

4. Liquid interface will be visible within 1 h. Allow the gel to polymerize for another 2 h at least. Rinse the surface of the gel with ddH_2O before pouring the stacking gel (4% stacking gel (10 mL): 1.35 mL acrylamide/bis-acrylamide solution, 2.5 mL stacking gel buffer, 50 μL 20% SDS, 100 μL 10% APS, 10 μL TEMED, and 6 mL ddH_2O). Insert the 15 -well comb taking care not to trap any bubbles below the teeth.
5. Add the laemmli loading buffer (1:1) to the sample (200 μg) and boil at 100 °C for 3–5 min. Allow the sample to cool at room temperature until loading.
6. Load samples and run gel at 25 mA in 1× SDS running buffer.
7. At this point, the gel can be stained with Coomassie blue (see below).
8. Place gel in a container. Cover with staining solution and shake at room temperature for 4 h.
9. Pour off staining solution. Add destaining solution to Coomassie blue stained gel and shake at room temperature for 2 h.
10. Decant the destaining solution.
11. Repeat **step 10** till bands become visible and no background could be observed in the gel.
12. Keep gel in water at room temperature. Digitize the gel image (Fig. 2d).

3.6 2-DE

The authors' laboratory performs 2-DE according to standard procedure with few modifications. The entire protocol is described in brief.

1. Perform isoelectric focusing (IEF) on 13 cm IPG strips with 250 μg of protein for pH range 4–7 13 cm and 1000 μg of protein for pH 4–7 24 cm strips (*see* **Notes 24–28** and Fig. 3a–c).
2. For pH gradient strip 4–7 13 cm and 24 cm, dilute aliquots of proteins with IEF rehydration buffer (*see* **Note 26**) and load 250 μL and 450 μL, respectively solution by the in-gel rehydration method (*see* **Notes 25** and **27**).
3. Perform IEF at 20 °C up to 42,000 and 78,000 Vh for 4–7 13 cm and 24 cm IPG strips, respectively (*see* **Notes 29** and **30**).
4. For reduction, treat the focused strips with 1% (w/v) DTT in 10 mL and 15 mL of equilibration buffer (*see* **Note 31**) for 20 min and 30 min, respectively for 4–7 13 cm and 24 cm IPG strips.
5. Treat the strips with 2.5% (w/v) iodoacetamide in equilibration buffer for another 20 min and 30 min, respectively for 4–7 13 cm and 24 cm IPG strips.

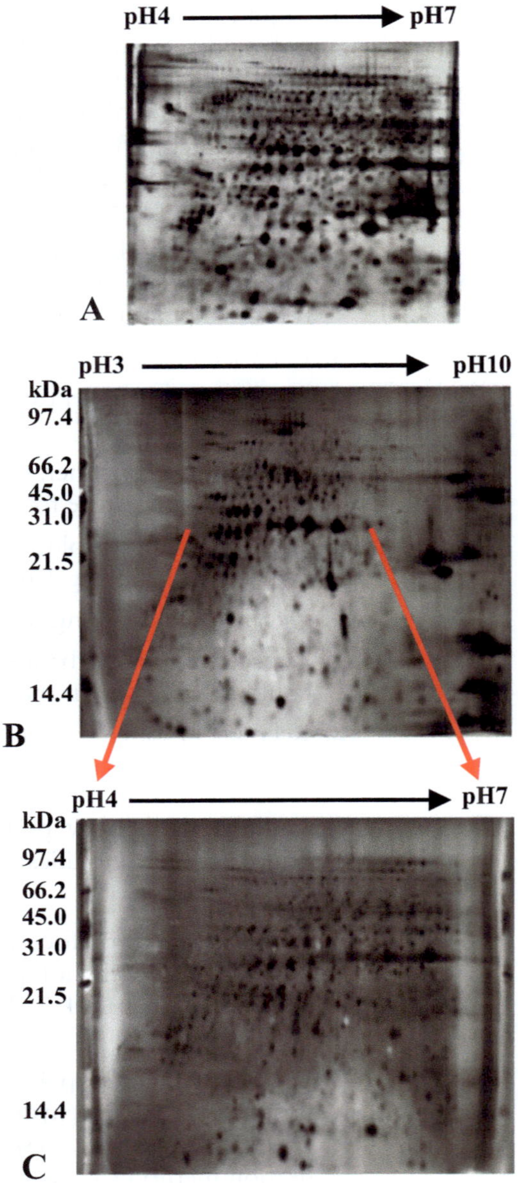

Fig. 3 Different Resolution of ECM proteins on 2-DE separated onto 12.5% SDS-PAGE and stained with silver stain. ECM proteins were electrofocused and zoomed onto (**a**) 13 cm IPG strip (pH 4–7); (**b**) 24 cm IPG strip (pH 3–10) and (**c**) 24 cm IPG strip (pH 4–7)

6. Dip the strips in 1× running buffer and then load on the top of 12.5% polyacrylamide gels for SDS-PAGE. Position the IPG strips between the plates on the surface of the second dimension gel with the plastic backing against one of the glass plates

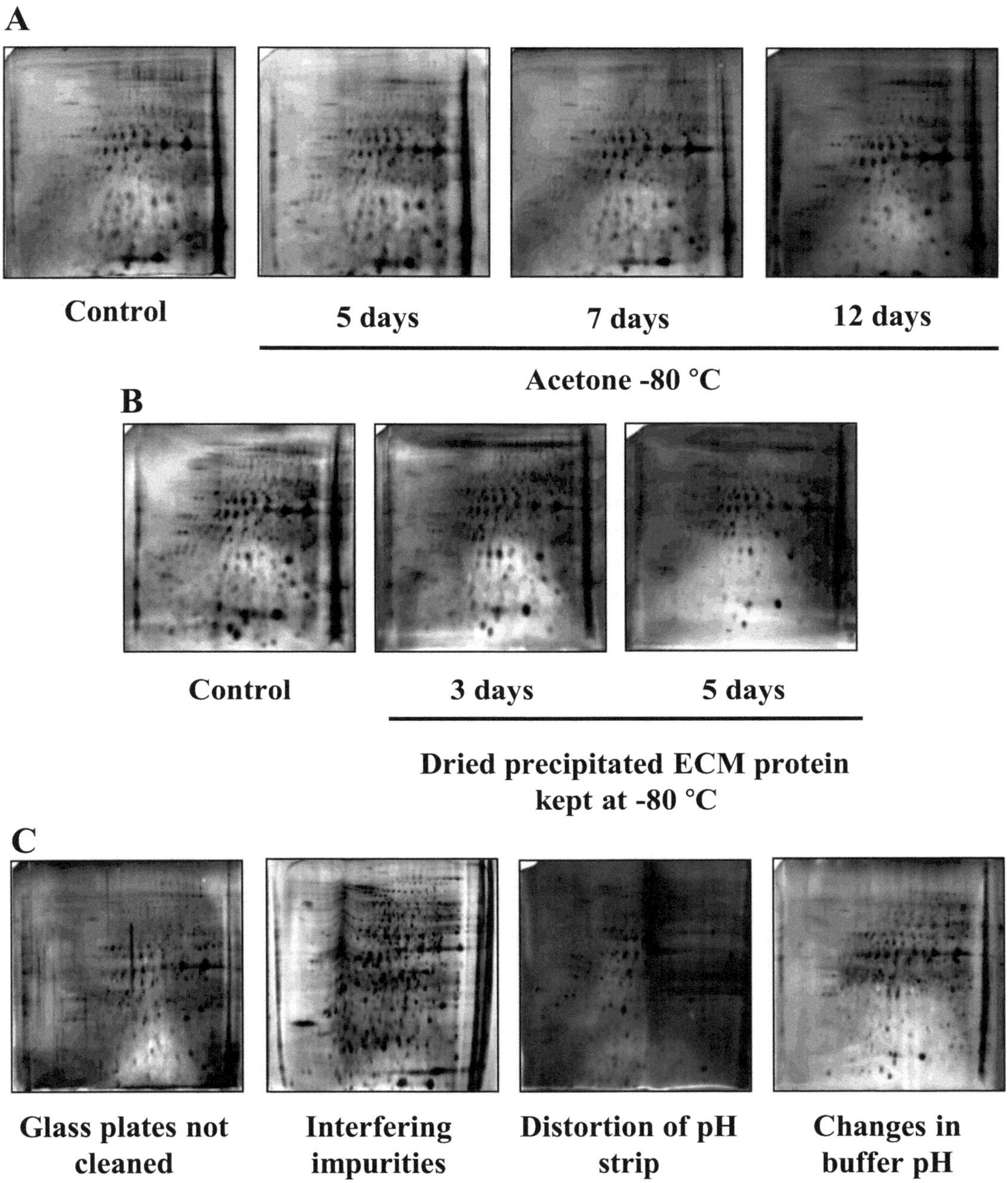

Fig. 4 Representative 2-DE gels. Stability of ECM proteins kept in −80 °C over a period of 12 days (**a**) with acetone, (**b**) dry pellet, then separated onto 13 cm strip p*I* 4–7 and stained with silver, (**c**) causes of improper resolution of ECM proteins on 2-DE gels

(*see* **Note 32**). Add the molecular weight markers to Whatman paper in a volume of 5 μL and apply to the top of the gel at one end of the IPG strip. Seal the strips in place using the agarose sealing gel.

7. Perform electrophoresis at constant current in two steps, initially fixed at 15 mA and 25 mA per gel (during transfer of protein from IPGstrip to second dimension gel), respectively for 13 cm and 24 cm IPG strips and then 30 mA and 70 mA per gel, respectively for 13 cm and 24 cm IPG strips till the dye front reaches approximately 1 mm from the bottom of the gel (*see* **Note 33**).
8. Disassemble the gel stacks. Electrophoresed proteins can be stained with Silver Stain Plus kit (Bio-Rad) as per the manufacturer's instructions. Gel images can be digitized with a Bio-Rad FluorS equipped with a 12-bit camera (*see* **Notes 34** and **35**, Fig. 4).

4 Notes

1. Seedlings ground after they are frozen and thawed shows a slightly different protein profile than those that are ground fresh or those that are ground while frozen. Decide which method works best. Consistency while grinding is important.
2. The yield of ECM protein for that amount of seedling should be more than 5 mg.
3. Grinding with a mortar and pestle rather than disrupting with a homogenizer yielded more reproducible results in our experiments.
4. All glasswares should be sterilized by autoclaving at 121 °C under 15 lb. psi pressures for 15 min.
5. PMSF is unstable in aqueous solutions with a half-life of approximately 30 min. Add solution immediately before use.
6. Prepare all the solutions in ultrapure water and analytical grade reagents.
7. Store all reagents at room temperature (unless indicated otherwise).
8. Acrylamide and bis-acrylamide are neurotoxic in nature. All the steps should be performed wearing powder-free gloves.
9. In acrylamide solution mixed resin AG 501-X8 (D) (anion and cation exchange resin) removes charged ions (e.g., free radicals) and allows long storage. Upon exhaustion of the exchange capacity, the color of resin changes from blue–green to gold. The used mixed resin can be filtered out before storage at 4 °C.
10. Initially add only Tris and glycine to 800 mL of water. Let it dissolve and then add SDS and make up the volume.
11. Homogenization buffer for ECM isolation is supplemented with PVPP to absorb phenolic compounds.

12. Freezing of plant tissues may lead to disintegration of tissues and thus fresh biomaterials should be preferred. However, our protocol was equally effective with the use of frozen plant materials.
13. Ratio of homogenization buffer to tissue weight needs to be optimal as inadequate buffer amount may result in impure ECM preparations.
14. Higher speed centrifugation would form a firmer pellet, but it would also pull proteins out of the supernatant. Precaution: Don't increase the centrifugal speed beyond $1000 \times g$ to prevent sedimentation of other cellular components.
15. Perform sufficient volume and number of washes to ensure clean ECM. Recommended ratios are approx. 5 mL/g) that for each H_2O wash, and ten times for acetone wash.
16. Keep samples at low temperature during extraction and extracted protein samples should be kept at 22–25 °C to avoid urea precipitation.
17. Acetone precipitates proteins, disrupts membranes and allows the removal of membrane proteins, chlorophyll, and other contaminating substances.
18. Protein yield is higher if the ECM is extracted overnight in the cold, but a reasonable protein yield can still be obtained with a shorter extraction (2 h).
19. Cell wall protein fraction is stable when kept in acetone at −80 °C for 10 days while degradation will be there if the cell wall protein fraction was kept as dry pellet in −80 °C.
20. Aliquots of lysis and rehydration buffers are stored indefinitely at −20 °C.
21. Marker assays should be carried out in triplicate to evaluate possible contamination.
22. The assay was performed in the presence of a detergent to expose all hidden sites for the cell wall bound plasma membrane, if any.
23. Use of dirty plates gel casting or presence of particulate material on the surface of the gel leads to streaking and change in spot morphology.
24. Protein samples containing Interfering substances or nonprotein impurities can interfere with IEF causing horizontal streaking.
25. Always use the recommended amount of rehydration buffer depending on the length of IPG strips and ensure that there are no bubbles between the rehydration solution and the gel surface.

26. Never heat urea solutions above room temperature. Heating converts urea into isocyanic acid that carbamylate amino acids resulting in charge shifts during isoelectric focusing.
27. Profile changes as pH strip touched surface while isofocusing.
28. Rehydrated IPG strips should always be under IPG cover fluid to prevent strip dehydration that would lead to improper focusing.
29. As isoelectric focusing proceeds, the bromophenol blue tracking dye migrates toward the anode. If the dye does not migrate, it indicates that no current is flowing. In this situation, check the contact between the strip holder electrodes and the electrode areas on the IEF system, and between the rehydrated gel and the internal face of the electrode.
30. Avoid both underfocusing and overfocusing.
31. Protein spot distribution and profile changes with change in buffer pH.
32. Entire lower edge of the IPG strip should be in contact with the top surface of the second dimension gel for effective protein transfer. No air bubbles should be trapped between the IPG strip and the gel surface or between the gel backing and the glass plate.
33. During second dimension electrophoresis it is important that add the buffer slowly to prevent bubbles formation on lower gel edge. We typically do not add the buffers until samples are ready to load to minimize diffusion of buffers between gel and tank.
34. Avoid dehydration of the gels by keeping it in water or 1% acetic acid solution for several days.
35. If TCA/acetone or methanol is used the following problems would arise:
    (a) Difficulty in isoelectric focusing (IEF). Horizontal and vertical streaking in 2-DE gels due to polysaccharide or other contaminants in TCA/acetone precipitated ECM protein samples.
    (b) Low resolution of proteins in methanol-precipitated samples.

Acknowledgments

This work was supported by grants from National Institute of Plant Genome Research, New Delhi, India and Department of Biotechnology, Govt. of India (No. BT/PR10796/BRB/10/621/2008, BT/HRD/35/01/05/2013 and BT/PR23748/BPA/118/345/2017) to S.C. E.E. was the recipient of postdoctoral fellowship from DBT-TWAS. K.N. is the recipient of postdoctoral fellowship from Department of Biotechnology (DBT), Govt. of India. The authors also thank Jasbeer Singh for illustrations and graphical representations in the manuscript.

References

1. Lilley KS, Dupree P (2007) Plant organelle proteomics. Curr Opin Plant Biol 10:594–599
2. Ghimire B, Riley WJ, Koven CD et al (2017) A global trait-based approach to estimate leaf nitrogen functional allocation from observations. Ecol Appl 27:1421–1434
3. Lambers H, Poorter H (1992) Inherent variation in growth rate between higher plants: a search for physiological causes and ecological consequences. Adv Ecol Res 23:187–261
4. Lamport DTA (1965) The protein component of primary cell walls. Adv Bot Res 2:151–218
5. Narula K, Elagamey E, Datta A et al (2012) Comparative analyses of extracellular matrix proteome: an under-explored area in plant research. In: Goyal A (ed) Crop plants. InTech, Janeza Tradine, Croatia
6. Landi S, Esposito S (2017) Nitrate uptake affects cell wall synthesis and modeling. Front Plant Sci 8:1376. https://doi.org/10.3389/fpls.2017.01376. eCollection 2017
7. Elagamey E, Narula K, Sinha A et al (2017) Quantitative extracellular matrix proteomics suggests cell wall reprogramming in host-specific immunity during vascular wilt caused by Fusarium oxysporum in chickpea. Proteomics 17:1600374
8. Robertson D, Mitchell GP, Gilroy JS et al (1997) Differential extraction and protein sequencing reveals major differences in patterns of primary cell wall proteins from plants. J Biol Chem 272:15841–15848
9. Nguyen-Kim H, San Clemente H, Balliau T et al (2016) Arabidopsis thaliana root cell wall proteomics: increasing the proteome coverage using a combinatorial peptide ligand library and description of unexpected Hyp in peroxidase amino acid sequences. Proteomics 16 (3):491–503
10. Albenne C, Canut H, Jamet E (2013) Plant cell wall proteomics: the leadership of Arabidopsis thaliana. Front Plant Sci 4:111
11. Smith JJ, Muldoon EP, Lamport DTA (1984) Isolation of extension precursors by direct elution of intact tomato cell suspension cultures. Phytochemistry 23:1233–1239
12. Borderies G, Jamet E, Lafitte C et al (2003) Proteomics of loosely bound cell wall proteins of Arabidopsis thaliana cell suspension cultures: a critical analysis. Electrophoresis 24:3421–3432
13. Lee SJ, Saravanan RS, Damasceno CMB et al (2004) Digging deeper into the plant cell wall proteome. Plant Physiol Biochem 42:979–988
14. Bayer E, Thomas CL, Maule AJ (2004) Plasmodesmata in Arabidopsis thaliana suspension cells. Protoplasma 223:93–102
15. Chivasa S, Ndimba BK, Simon WJ et al (2002) Proteomic analysis of the *Arabidopsis thaliana* cell wall. Electrophoresis 23:1754–1765
16. Ndimba BK, Chivasa S, Hamilton JM et al (2003) Proteomic analysis of changes in the extracellular matrix of Arabidopsis cell suspension cultures induced by fungal elicitors. Proteomics 3:1047–1059
17. Zhu J, Chen S, Alvarez S et al (2006) Cell wall proteome in the maize primary root elongation zone. I. Extraction and identification of water-soluble and lightly ionically bound proteins. Plant Physiol 140:311–325
18. Kärkönen A, Koutaniemi S, Mustonen M et al (2002) Lignification related enzymes in Picea abies suspension cultures. Physiol Plant 114:343–353
19. Irshad M, Canut H, Borderies G et al (2008) A new picture of cell wall protein dynamics in elongating cells of Arabidopsis thaliana: confirmed actors and newcomers. BMC Plant Biol 8:94
20. Jamet E, Albenne C, Boudart G (2008) Recent advances in plant cell wall proteomics. Proteomics 8:893–908
21. Bhushan D, Pandey A, Chattopadhyay A et al (2006) Extracellular matrix proteome of chickpea (Cicer arietinum) illustrates pathway abundance, novel protein functions and evolutionary perspect. J Proteome Res 5:1711–1720
22. Bhushan D, Pandey A, Choudhary MK et al (2007) Comparative proteomics analysis of differentially expressed proteins in chickpea extracellular matrix during dehydration stress. Mol Cell Proteomics 6:1868–1884
23. Pandey A, Rajamani U, Verma J et al (2010) Identification of extracellular matrix proteins of rice (Oryza sativa L.) involved in dehydration-

responsive network: a proteomic approach. J Proteome Res 9:3443–3464

24. Bhushan D, Jaiswal DK, Ray D et al (2011) Dehydration-responsive reversible and irreversible changes in the extracellular matrix: comparative proteomics of chickpea genotypes with contrasting tolerance. J Proteome Res 10:2027–2046
25. Elagamey E, Narula K, Sinha A et al (2016) Extracellular matrix proteome and phosphoproteome of potato reveals functionally distinct and diverse canonical and non-canonical proteoforms. Proteomes 4:20
26. Elagamey E, Sinha A, Narula K et al (2017) Molecular dissection of extracellular matrix proteome reveals discrete mechanism regulating Verticillium dahliae triggered vascular wilt disease in potato. Proteomics 17:1600373

Index

Kapuganti Jagadis Gupta (ed.), *Nitrogen Metabolism in Plants: Methods and Protocols*, Methods in Molecular Biology, vol. 2057, https://doi.org/10.1007/978-1-4939-9790-9, © Springer Science+Business Media, LLC, part of Springer Nature 2020

D

E

F

G

H

I

J

K

L

M

N

O